Molecular Pathology

APPROACHES TO DIAGNOSING HUMAN DISEASE IN THE CLINICAL LABORATORY

Molecular Pathology

APPROACHES TO DIAGNOSING HUMAN DISEASE IN THE CLINICAL LABORATORY

Ruth A. Heim
Lawrence M. Silverman

CAROLINA ACADEMIC PRESS

ISBN: 0-89089-544-9
LCCN: 93-74307

Carolina Academic Press
700 Kent Street
Durham, North Carolina 27701
(919) 489-7486
FAX (919) 493-5668

Printed in the United States of America

Over the past decade, many fellows, residents, technologists, and students have greatly contributed to the development of our laboratory and our training program. This book is dedicated to them.

Contributors

Nancy P. Callanan, M.S.
Division of Genetics and Metabolism
Department of Pediatrics
University of North Carolina
Chapel Hill, NC 27599

Georgette A. Dent, M.D.
Hematology Laboratory
Department of Hospital Laboratories
University of North Carolina Hospitals
Chapel Hill, NC 27514

Joseph J. Eron, M.D.
Infectious Disease Laboratory
Department of Medicine
University of North Carolina
Chapel Hill, NC 27599

Susan A. Fiscus, Ph.D.
Retrovirology Laboratory
Department of Hospital Laboratories
University of North Carolina Hospitals
Chapel Hill, NC 27514

Kenneth J. Friedman, B.A.
Molecular Genetics Laboratory
Department of Hospital Laboratories
University of North Carolina Hospitals
Chapel Hill, NC 27514

Ruth A. Heim, Ph.D.
Molecular Genetics Laboratory
Department of Hospital Laboratories
University of North Carolina Hospitals
Chapel Hill, NC 27514

W. Edward Highsmith, Ph.D.
Applied Technology Genetics
Corporation
30 Spring Mill Drive
Malvern, PA 19355

Roy L. Hopfer, Ph.D.
Clinical Microbiology-Immunology
Laboratories
Department of Hospital Laboratories
University of North Carolina Hospitals
Chapel Hill, NC 27514

Thomas W. Prior, Ph.D.
Department of Pathology, Ohio State
University
121 Hamilton Hall, 1545 Neil Avenue
Columbus, OH 43210

Lawrence M. Silverman Ph.D.
Molecular Genetics Laboratory
Department of Hospital Laboratories
University of North Carolina Hospitals
Chapel Hill, NC 27514

Gregory J. Tsongalis, Ph.D.
Molecular Genetics Laboratory
Department of Hospital Laboratories
University of North Carolina Hospitals
Chapel Hill, NC 27514

Emily A. Winn-Deen, Ph.D.
Applied Biosystems Division of Perkin
Elmer Coporation
850 Lincoln Center Drive
Foster City, CA 94404

Contents

Color Illustrations follow page 56.

Foreword

The emerging field of Molecular Pathology, the subject of this timely volume, can be defined as a discipline in which clinical inferences are made from the analysis of nucleotide sequence information derived from the clinical specimen, in accord with the new molecular injunction, "by their nucleotide sequences, so shall ye know them." The identification of infectious disease pathogens, genetic disease diagnosis and carrier screening, and, in some cases prognosis for specific cancers can be carried out by detecting specific DNA sequences. Two related developments have given rise to this new and rapidly growing field. First, stimulated by the development of recombinant DNA/molecular cloning and sequencing techniques in the mid-70s, the genomes of a variety of viruses, bacteria, and other micro-organisms were characterized; in addition, many human genetic diseases, such as the hemoglobinopathies (e.g. sickle-cell anemia and beta-thalassemia) were traced to one or to several specific mutations in specific genes (e.g. beta-globin).

The second major development has been a series of critical technological advances in methods for accessing clinically valuable information embedded in DNA sequences. The availability of a variety of restriction endonucleases to cut genomic DNA into fragments, gel electrophoretic methods to separate the fragments, and cloned gene segments to serve as specific labeled hybridization probes made the Southern blotting procedure a powerful method for detecting polymorphisms and mutation. Another important technical advance has been the transition from the use of radioactive to non-radioactive "reporter' molecules, such as enzymes with chromogenic or chemiluminescent substrates, to label DNA hybridization probes. This elimination of radioisotopes has been critical to the acceptance of these powerful molecular biology tools in a clinical diagnostic setting.

Perhaps, the most dramatic of these advances has been the development in the mid-1980's of the polymerase chain reaction (PCR), a method for synthesizing millions of copies of a specific DNA segment from a complex template. Following the introduction of a thermostable DNA polymerase isolated from the thermophilic bacterium, *Thermus aquaticus* and the availability of automated thermal cycling instruments, the capacity of PCR to amplify specific DNA sequences from virtually any kind of specimen created a revolution in molecular biology procedures in basic research as well as in clinical diagnostic labs. In the past several years, many advances in PCR technology, including the ability to amplify efficiently from an RNA template, following reverse transcription into cDNA, and the recently developed methods for *in situ* amplification of histological specimens on glass slides, have greatly expanded the range of clinical applications.

The ability of PCR to amplify and analyze the genotype of a single cell has permitted not only new approaches to genetic mapping but also clinical genetic analysis of pre-implantation embryos. This capacity for the genetic analysis of minute samples has proved extremely valuable in forensic analysis; PCR-based forensic genetic typing has been used in the U.S. since 1986. In a clinical pathology context, such genetic markers for individualization have proved useful for resolving sample mix-ups and for confirming that sequential specimens were drawn from the same individual. In addition, the analysis of pathogen sequence variation has proved an invaluable epidemiologic tool.

Since the first report of PCR in 1985, a variety of techniques, accompanied by a bewildering proliferation of acronyms (or BPA) have been developed for detecting and analyzing DNA amplified from clinical specimens. These methods are now being widely used in clinical reference laboratories and some have been incorporated in commercial diagnostic kits.

Some of the clinical procedures made possible by PCR amplification, such as pre-implantation genetic diagnosis, raise complex ethical and social issues for the molecular pathologist and for the genetics community in general, as does the identification of genes which confer increased susceptibility to specific diseases. Our current knowledge of clinically relevant DNA sequences will expand dramatically as a result of the Human Genome project and related research activities around the world. The capacity to obtain this information with increasingly simplified laboratory procedures poses great opportunities and challenges for molecular pathology. For molecular biologists and clinicians alike, we are entering a new and exciting era.

Henry A. Erlich
May 3, 1994
Roche Molecular Systems, Inc.
1145 Atlantic Avenue, Suite 100
Alameda, CA 94501

Preface

Perhaps the first description of clinical molecular genetics can be traced to David Weatherall's book, "The New Genetics and Clinical Practice", published in 1985, in which the author stated in the introduction: "During the last few years there have been remarkable advances in molecular biology. . . . (that) have been applied to the study of human genes, both in health and disease. It is now possible to define many diseases in terms of their *molecular pathology*, a level of diagnostic precision that would have been undreamed of 10 years ago." From this beginning, we coined the term "molecular pathology" to refer to the applications of nucleic acid probe technology in the diagnostic laboratory, a designation analagous to that for surgical, anatomic, or clinical pathology.

At the University of North Carolina Hospitals in 1985, those exciting developments in molecular biology were generating intense interest in applying molecular technologies to clinical pathology. Several areas in our clinical laboratories were ripe for development, including hematopathology and molecular genetics, a new initiative for the hospital laboratories. During the past decade, molecular diagnostic techniques have been applied in these and other clinical laboratories at our institution, including cytogenetics, microbiology, infectious diseases, oncology, and more recently, transfusion medicine (for Rh blood group typing).

As these areas of molecular pathology have grown, laboratory accreditation and personnel certification issues, as well as cost and allocation of limited laboratory resources, have had to be addressed. This book describes the experience of individuals currently or previously trained at our institution, providing an insight into the extent to which molecular biological techniques have extended diagnostic capabilities at one institution. The book presents a cross-section through current standards of laboratory practice and, to a limited extent, makes some predictions for the future of molecular pathology.

CHAPTER I

Introduction

Ruth A. Heim and Lawrence M. Silverman

The Discipline of "Molecular Pathology"

The original description by Miescher in 1871 of "nuclein" in pus cells, and the classic paper by Watson and Crick in 1953 ("The Helical Structure of DNA and its Genetic Implications"), demonstrates the fascination that the biochemical nature of heredity has for scientists. In the last forty years, progress in understanding the structure and function of DNA has led to technical advances in manipulating and studying the genetic material. These advances constitute the recombinant DNA technologies developed by molecular biologists studying biological function at the DNA level. Recombinant DNA techniques have been applied to tissues, cells, whole chromosomes and DNA sequences, and their application has led to a better understanding of the pathogenesis, classification and diagnosis of human disease. This evolving understanding of human molecular pathology has fundamentally affected the practice of clinical medicine: human disease can now be diagnosed at the DNA level.

There are large numbers of potentially infectious disease agents and a large predicted number of potentially aberrant human genes (100,000–150,000), some of which are now and all of which could one day be amenable to DNA diagnosis. A new subdvision within the field of pathology is thus emerging—molecular laboratory testing. We refer to "molecular pathology" as a discipline that complements anatomic, surgical, and clinical pathology. This book constitutes an effort to review current practices in molecular pathology at our present state of knowledge, setting the stage for future developments in disease prevention and early therapeutic intervention.

Approaches to Diagnosing Human Disease

Recombinant DNA procedures are being applied in clinical practice to identify molecular changes associated with neoplasia, heritable diseases, and acquired infectious diseases. These diagnostic applications are the subject of this book. We begin with a chapter (chapter 2) on the automation of molecular techniques used in DNA diagnostic testing, because the clinical application of diagnostic procedures depends on technical advances that improve simplicity, accuracy, speed and cost. Some of these advances include use of non-isotopic methods rather than those requiring radioisotopes, and decreased reliance on the labor-intensive and cost-ineffective hybridization techniques widely used in the last decade. One important rate-limiting factor in the clinical application of DNA diagnostic tests is the need for speedier and more effective mutation detection methods.

In chapters 3 and 4, molecular pathologists describe the application of recombinant DNA techniques to detecting not only acquired mutations in somatic cell DNA associated with neoplastic transformation, but also inherited mutations that predispose individuals to neoplastic transformation. Direct molecular tests can detect oncogenes activated by such mechanisms as point mutations (for example, in the p53 gene) or chromosomal rearrangements (for example, the bcr/abl translocation), or can detect somatic losses (for example, in retinoblastoma or Wilm's tumor) or amplifications of genetic sequences responsible for neoplastic phenotypes (for example, the HER/neu gene). It is now possible to determine if certain tumors are malignant or benign, or to decide the prognosis for a leukemic patient, using DNA diagnostic tests.

In chapters 5 to 9, molecular geneticists describe the diagnostic techniques available to provide direct and indirect tests for molecular defects associated with heritable diseases. DNA tests can provide genetic information to individuals either with symptoms of a disease or before symptoms are evident, as well as detect carriers of certain disease-associated genes or provide prenatal diagnosis. In chapter 5, the molecular biology of genetic diseases and approaches to their diagnosis are reviewed. The following three chapters focus on diagnosing specific diseases, each illustrating a different type of molecular defect and posing different diagnostic problems: point mutations in cystic fibrosis (chapter 6), large deletions in Duchenne and Becker muscular dystrophies (chapter 7), and trinucleotide repeat expansions in the fragile X syndrome and myotonic dystrophy (chapter 8).

In chapters 10 and 11, molecular virologists and microbiologists describe the contributions recombinant DNA techniques have made to the direct diagnosis of infectious disease agents. At present, the most extensive contributions have been in the detection of viral pathogens.

The Clinical Laboratory

DNA diagnostic tests are unusual in several respects. Unlike most other clinical tests, they do not measure any physiological state or visualize anatomy, and can be performed using fresh, frozen or archival tissue. Since the tests are specific to one disease, the results may not provide differential diagnostic information for another disease state, unlike an X-ray, for example. The tests themselves do not necessarily detect a specific mutation directly; many DNA tests for heritable diseases are indirect and based on prior diagnosis of the disease in other family members. For such tests samples may be required not only from the patient but also from one or more of the patient's blood-relatives.

The specificity of many DNA tests allows presymptomatic or prenatal diagnoses to be made or carrier status to be identified. Information of these kinds, unlike the information obtained from other clinical tests to be used by the physician to manage the patient's treatment or care, may be used primarily by the patient to make personal decisions about lifestyle and reproductive choices. The information obtained from DNA testing can thus have a profound effect on society. Counseling and social issues related to diagnosis of inherited diseases are discussed in chapter 9. Wide diagnostic application of recombinant DNA procedures will depend on a resolution of related medical, ethical and economic issues.

Lastly, DNA diagnostic tests are being used with other traditional tests, leading to changing interpretations of laboratory data. Until we have more experience with data from DNA tests, it will sometimes be difficult to establish the role of specific DNA-based information. A good example is in the field of cancer-related profiling. Tumor tissues are currently assessed by histology, immunohistochemistry, in situ hybridization, flow cytometry, image analysis, and other technologies. How some of the new DNA-based testing, including in situ amplification, will replace or complement conventional tumor analyses remains to be seen. Similarly, DNA-based information related to infectious diseases, such as the human immunodeficiency viruses, is currently being evaluated to determine clinical utility. Until these and more questions are answered, we must consider the development of molecular diagnostic tests as "work-in-progress." In the final chapter of this book, we selectively address scientific and related issues that have an impact on the future of molecular pathology.

CHAPTER 2

Molecular Techniques and Their Automation in the Clinical Laboratory

Emily S. Winn-Deen, Ph.D.

Introduction

Molecular techniques began as intensely manual methods requiring the use of specialized reagents, radioactive detection, and highly trained technical staff. This chapter contains a discussion of several of these techniques and how each has been automated to date. These include DNA purification, DNA sequencing, Southern blotting and the polymerase chain reaction (PCR). Over the last ten years automation has played an increasingly important role in making these techniques more robust and routine. In some cases automation has also been coupled with non-isotopic detection, making them more accessible to laboratories that do not wish to work with radioactivity. Each technique will be discussed in sufficient detail to give the reader a general overview of its clinical applications and the various approaches to its automation.

DNA Purification

The first step in any molecular assay is the isolation and purification of the DNA or RNA. Typically this nucleic acid is isolated from either blood or from cells such as amniocytes or buccal cells. When searching for infectious agents the samples may also include sputum, vaginal or urethral swabs, and stool. The type of sample greatly influences the way that the nucleic acid is isolated. For example the heme in red blood cells can interfere with DNA polymerase so that care must be taken in preparation of blood samples for PCR. Likewise, the restriction enzymes used for Southern blotting are very sensitive to impurities, and this application requires highly purified DNA.

Purification of nucleic acids begins with cell lysis (Sambrook *et al.* 1989). Mammalian cells can be lysed by boiling or treatment with a detergent such as sodium dodecyl sulfate (SDS). Cells from various microorganisms require lysis methods tailored to the characteristics of the particular organism. After lysis, many purification methods remove detrimental proteins such as nucleases through treatment with Proteinase K in the presence of EDTA and extraction with phenol. After removal of the protein, the nucleic acid is generally alcohol precipitated to yield high molecular weight DNA. These purification methods are a relatively straightforward process to do manually, but do involve several separation, centrifugation and precipitation steps where errors can occur. Samples can be mislabeled, dropped or cross-contaminated, and pipetting errors can be made at any of the many steps. Some samples such as human blood/tissue may contain infectious agents and therefore require special precautions for handling. Additionally, the chemicals used in the standard phenol/chloroform extraction procedure are hazardous.

The first approach to automation of this process was introduced by Applied Biosystems (Foster City, CA) in 1986. The GENEPURE™ 341 Nucleic Acid Purification System features on-line chemistry reagents and a unique reaction vessel available in three volumes (7, 14 and 30 mL) which can be heated and rotated. In a typical protocol, samples from blood, tissue, cell culture, plants, plasmids, bacteria or viruses are added to the reaction vessel. Hot lysis buffer, containing Proteinase K to cleave the proteins and deactivate nucleases, is added and gentle agitation is used to open the cells. After lysis is complete, phenol/chloroform is added, and the sample is extracted using gentle agitation. The reaction vessel returns to a horizontal position and phase separation is encouraged by the addition of heat. The lower layer (organic) is transferred to waste until the in-line conductivity meter senses the aqueous layer. Phenol extraction can be repeated if desired. The DNA is then precipitated by the addition of ethanol/sodium acetate and collected onto a special filter cartridge. After additional ethanol washing, the filter cartridge is removed and the DNA is redissolved in the appropriate buffer. DNA purified in this manner is suitable for most molecular biology applications including sequencing, polymerase chain reaction (PCR), and restriction digestion. Alternate protocols for RNA purification are also available. Not all end-uses require complete DNA purification and for them a fast, non-organic cycle can be used. This instrument can process eight samples to purified DNA in 2–4 hours, depending on the cycle chosen.

In 1990 AutoGen (Integrated Separation Systems, Natick, MA) introduced an alternative approach to automation of nucleic acid purification. This instrument provides a more conventional approach to automation of the manual chemistry. The operator loads 1 mL of cultured cells or blood into a microcentrifuge tube and places it into the centrifuge module. Empty tubes and pipette tips are also loaded into the unit and the cycle begins. Microprocessor control allows all subsequent reagent deliveries, aspirations and centrifugations to be handled by the instrument. These steps include cell concentration, lysis, neutralization, deproteination, precipitation, DNA washing and resuspension. The Model 540 can process up to 12 samples to purified DNA in 40 minutes. Protocols are available for isolation of DNA or RNA from blood, plasmids, cosmids, phages, plants cells, tissue suspensions, and bacterial cultures.

Recently the Beckman (Palo Alto, CA) Biomek® 1000 workstation, a generic X, Y, Z pipetting robot, has also been adapted to isolation of genomic DNA (Mischiati et al. 1993). The method involves automation of DNA isolation by anion exchange chromatography on Qiagen-100 columns (Chatsworth, CA). The robot handles all the liquid manipulations including loading cells onto the column, washing the column, and eluting the DNA. Using this method DNA from eight to twelve samples of $\sim 10^6$ cells each can be isolated in less than one hour. DNA prepared in this way is suitable for PCR or for allele-specific hybridization applications.

DNA Sequencing

DNA sequencing has traditionally been a research tool. As more genes are discovered and their sequence uncovered, sequencing is emerging as a powerful diagnostic tool as well. Diagnostic sequencing has been applied to a variety of problems, including HLA typing (Kaneoka et al. 1991; Santamaria et al. 1992), diagnosis of inherited diseases caused by point mutations in the hypoxanthine phosphoribosyltransferase (Gibbs et al. 1989, 1990) and apolipoprotein B genes (Leren et al. 1993), p53 tumor suppressor gene mutations (Kovach *et al.* 1991), and epidemiology of infectious diseases, such as cholera (Olsvik *et al.* 1993) and HIV (Wike *et al.* 1992; Wolinsky *et al.* 1992). This tool is the ultimate molecular gold standard against which any other method must be compared.

DNA Sequencing Using Radioactive Labels

In the mid 1970s a method of DNA sequence analysis based on chemical degradation of DNA was described (Maxam and Gilbert 1977). Later that year an enzymatic method based on random chain termination with dideoxynucleotides was reported (Sanger *et al.* 1977). A description of the Sanger sequencing method is shown in Figure 2.1. The original development of these DNA sequence analysis methods was done using radioactive markers such as ^{32}P or ^{35}S. For the Sanger method, a reaction mix containing a primer, DNA polymerase, and the four deoxynucleotide triphosphates (at least one of which is radioactively labeled) is prepared and then divided into four aliquots. A limiting amount of either A, T, C, or G dideoxynucleotide triphosphate is added to each of the four aliquots so that the dideoxy-A tube ends up containing all of the fragments which contain terminal A, the dideoxy-T tube ends up containing all of the fragments which contain terminal T, and so on. After the sequencing reactions are completed the A, T, C, and G aliquots are each loaded onto a separate lane of a denaturing acrylamide gel and separated by high voltage electrophoresis. The gel is then transferred to a solid support and subjected to autoradiography. The sequence is read by following the bands between the A, T, C, and G lanes. This method offers a relatively simple means of analyzing DNA sequences and was quickly adopted by the research community. However, the pipetting and transfer required are quite tedious, and accurate interpretation of base sequence from the autoradiograms requires skill and careful attention to detail.

Automation for radioactive DNA sequencing has taken several forms. In order to eliminate the need to transfer after electrophoresis, two instruments are now available which combine electrophoresis and transfer into a single operation (Intelligenetics/Betagen AutoTrans System, Mountain View, CA; Hoefer TE 2000 TwoStep™, San Francisco, CA). Both electrophorese a sample down a vertical gel and, as the DNA elutes off the bottom of the gel, it is deposited onto a constantly moving membrane web. With the AutoTrans System a computer controls web speed, electrophoresis voltage, temperature, and buffer recirculation, while the TwoStep™ controls only web speed and temperature.

Another part of the radioactive sequencing process that has been chosen for automation is the creation and analysis of the autoradiogram. Molecular Dynamics (Sunnyvale, CA) has a computing densitometer which can be used to measure the intensity of the bands from the exposed X-ray film and their ImageQuant™ software can then be used to facilitate interpretation of this data. Similar commercially available systems are the Discovery Series™ (protein and DNA imageWare systems, Huntington Station, NY) and the LYNX densitometer (Applied Imaging, Santa Clara, CA). Other companies have automated the actual radiation measurement by replacing traditional autoradiography with direct reading of the radioac-

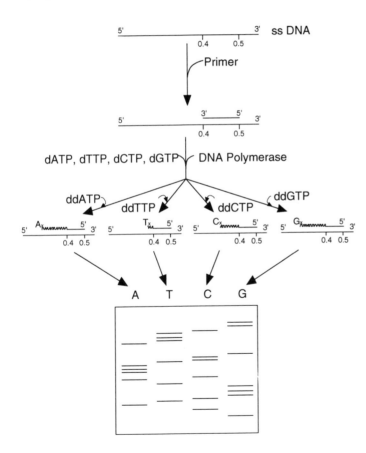

Figure 2.1 Schematic of the Sanger dideoxy sequencing method (artwork courtesy of Ellson Chen, Applied Biosystems).

tive blot. The Intelligenetics/Betagen Betascope 603 Blot Analyzer™ images patterns of radioactivity from ^{32}P- or ^{35}S-labeled blots by tracing the path of high energy beta emission through two measurement planes, each consisting of three frames of tightly spaced wires, within a gas filled chamber. The instrument provides real time visualization of the radioactive areas of the blot and computer analysis of the counts coming from each area. This system is simple and easy to use. Its one drawback is that for samples which are weakly labeled the instrument may be tied up for many hours counting a single blot.

This problem is addressed by the Molecular Dynamics PhosphorImager. With this instrument each gel transfer is placed in a cassette which contains a BaFBr:Eu^{+2} medium to capture all the energy released by beta particles (Johnson et al. 1990). The beta particles excite an electron of the Eu^{+2}*, ion. This electron is then trapped in the BaFBr- complex and results in the oxidation of Eu^{+2} to Eu^{+3}. After exposure is complete, the phosphor screens are placed in the PhosphorImager and 633 nm light from a helium-neon laser converts the Eu^{+3} to an excited state, Eu^{+2}*, which in turn emits a photon at 390 nm as it returns to the ground state. These photons are counted by the PhosphorImager and their locations determined. This medium has a greater dynamic range than conventional X-ray film and requires much shorter exposure times. Since the instrument is used only for reading the medium, multiple gels can be exposing the phosphor plates at the same time in a manner analogous to autoradiography.

Automation has also been applied to analysis of the autoradiograms in order to determine base sequence. Variation in band mobility between lanes, and lanes which do not run straight down the gel, make radioactive sequencing difficult to interpret electronically. Chemistry anomalies such as compressions and stops also make instrument interpretation difficult. Several companies have written highly sophisticated software which interacts with a densitometric film scanner to try to address this problem. This software can find lanes, detect bands, correct for smiling and read out the sequence. Scanner/software packages are available from Millipore Corporation, Bedford, MA (Bio Image® 60S), Molecular Dynamics, Sunnyvale, CA (DNAscan™ software for use with the ImageQuant™ workstation, 325 Computing Densitometer, or PhosphorImager), and United States Biochemical, Cleveland, OH (SciScan™ 5000 with BioAnalysis™ software). These systems can significantly improve on the eyestrain associated with

manual interpretation of sequencing autorads, but provide limited automation of the sequencing process, as the gels must still be run, transferred
and exposed to film before interpretation can begin.

DNA Sequencing Using Fluorescent Labels

Several years after the first description of the Sanger enzymatic sequencing chemistry, a group of researchers at the California Institute of Technology and Applied Biosystems developed a set of fluorescent labels which
were to revolutionize DNA sequencing (Smith et al. 1985, 1986). These
dyes allowed the substitution of fluorescence for radioactivity and paved the
way for complete automation of DNA sequencing.

The first fluorescent-based DNA sequencer was introduced by Applied
Biosystems in 1986. It utilizes Sanger sequencing reactions run exactly as
they are for radioactive sequencing except that a fluorescently labeled
primer is substituted for the radioactive label. The A reactions contain a
primer labeled with one dye, while the T, C and G reactions use the same
primer sequence, but each labeled with a different rhodamine or fluorescein dye. The dyes are chosen such that they are all excited by the argon
ion laser used in the instrument, but emit at wavelengths which are approximately 20 nm apart (535 nm, 555 nm, 580 nm, and 605 nm). The overlap in the fluorescence emission spectra of the dyes requires software which
is capable of doing multicomponent analysis to eliminate the bleed-through
of one dye into another. By using special algorithms developed to handle
this problem, the colored bands are clearly resolved. The variation in electrophoretic mobility caused by the differing dye structures is controlled
by insertion of a variable length spacer arm between the dye and the 5' end
of the primer, decreasing relative mobility shifts to approximately one-
quarter of a base (Connell et al. 1987).

After the sequencing reactions are complete, the A, T, C, and G tubes
are pooled together and the DNA is ethanol processed. The DNA is then
resuspended in loading buffer and loaded onto an acrylamide gel. Due to
the unique emission spectra of each of the four dyes, the A, T, C and G samples can be co-electrophoresed in a single lane. Each fragment's fluorescence
is excited by directing the argon laser through a small area in the glass
plates near the bottom of the gel. Light emitted from the dye-labeled DNA
fragments passing through this area is focused by a collection lens through
a four wavelength selectable filter (540 nm, 560 nm, 580 nm, 610 nm) into

a photomultiplier tube (PMT), and digitized signals from the PMT are then transferred to the computer for interpretation. Software is also used to analyze emission wavelength, peak height, shape, and interval to provide automatic base calling. This approach allows up to 36 A, T, C, G sets to be run simultaneously on the instrument. Currently 700 bases of sequence information can be obtained per lane with 98% accuracy, corresponding to 25,200 bases of sequence information per gel. The advantages of this multi-color approach to sequencing are the elimination of the lane-to-lane drift associated with classical radioactive sequencing and a four-fold increase in the through-put per gel. A comparison of four-lane sequencing and one-lane sequencing is shown in Figure 2.2 (located in the color section following page 56).

In 1987 DuPont (Wilmington, DE) scientists took the concept of fluorescent sequencing one step further by placing the fluorescent label on the dideoxyterminator instead of on the primer (Prober et al. 1987). Four modified fluorescein dyes were chosen to minimize shifts in relative mobility between the four bases, while still allowing differentiation of the four dyes from their emission spectra. This approach has several advantages. First, it eliminates the need to synthesize fluorescently labeled primers for each sequence one wishes to analyze. Although dye-labeled primers for most of the common sequencing vectors are commercially available and custom dye-labeled primers can be synthesized, some sequencing applications such as gene walking are slowed down by the need to synthesize dye-labeled sequencing primers for each walk one wishes to take. Dye terminators also allow the A, T, C, and G reactions to take place in a single tube, thus cutting the chemistry reagents by a factor of four. A disadvantage of this method remains that excess dye-terminators must be removed on a spin column prior to loading the sample on the gel. However, by placing the dye on the terminator, signals from false terminations do not interfere with reading the sequence. Another problem which remains to be solved is efficient incorporation of dideoxy terminators by a variety of sequencing enzymes. The Klenow fragment of DNA polymerase performs poorly, and even with T7 polymerase and Taq polymerase certain sequences still prove difficult. A combination of dye-primer and dye-terminator sequencing is probably the best approach to complex sequencing problems. Dye-labeled terminators are available from Applied Biosystems.

Other fluorescent sequencers are also available. A system based on the use of a single fluorescein-labeled primer was developed in 1989 by scientists at the European Molecular Biology Laboratory (Ansorge *et al.* 1989) and is available from Pharmacia (Piscataway, NJ) as the Automated Laser Fluorescent (A.L.F.) DNA Sequencer™. Four separate reactions (A, T, C, G) must be run for each sequence to be determined and, as with radioactive sequencing, each of the four reactions must still be loaded into a different lane of the gel and electrophoresed side-by-side. The A.L.F. system's optics are simpler than the Applied Biosystems sequencer. The excitation laser is stationary and its light enters through the side of the gel. The 40 lanes of the gel are each equipped with a separate fixed detector to collect the emitted light. In addition to no moving optical parts, the system's laser light passes directly into the gel, thus minimizing problems with fluorescence of the glass plates. This system can determine about 350 bases for 10 sequences simultaneously (3500 bases/gel). However, it still suffers from the main problems which plague conventional radioactive sequencing, variation in electrophoretic mobility between lanes and its effects on accurate base-calling. Some of this is addressed by software which analyzes four lanes at once to assure that only one peak is found for any given base position. However, fixed detectors make it difficult to compensate for lanes which do run straight down the gel.

A recent entry into the fluorescent sequencer market is LI-COR (Lincoln, NE) (Middendorf *et al.* 1992). Their Model 4000 sequencer uses an infrared fluorescent dye-labeled primer and a inexpensive red laser similar to those used to read compact discs. The 10 mW solid state laser diode used for excitation at 785 nm and the confocal microscope objective/avalanche photodiode detector at 820 nm used for emission detection have allowed miniaturization of the instrument optics. This in turn has allowed the complete optics package to be mounted onto the scanning platform, giving flexibility in scan speed and scan width. Like the A.L.F., this instrument uses a single dye, and A, C, G, and T reactions must be run in separate lanes. Up to 11 samples can be run simultaneously (44 lanes). Data is analyzed using the system's Base ImagIR™ software with typical performance of 450 bases per sample with 99% accuracy.

Another approach reported recently (Ishino *et al.* 1992) involves using rhodamine X-labeled fluorescent primers to run the sequencing reactions. After running the samples out on a gel, the gel and plates are placed into a fluorescent image scanner and excited with a 15 mW green laser at 532 nm. The emitted light is monitored at 605 nm and an image created. This hybrid technique allows the user to review the sequence data quality quickly and return the gel to the electrophoresis chamber for further separation if initial resolution is not sufficient.

Robotic Preparation of Sequencing Reactions

With the advent of automated fluorescent sequencers came the need to automate the sequencing chemistries run prior to gel electrophoresis. Accurate, repetitive pipetting of small volumes in multiple reaction tubes is tedious and provides numerous opportunities for operator error. Two companies have provided a solution to this problem. The Beckman Biomek® 1000 workstation can be adapted through the purchase of specific modules to perform DNA sequencing (Wilson et al. 1988; Civetello et al. 1992). Applied Biosystems' Catalyst™ is a dedicated robotic workstation designed specifically for sequencing applications (Cathcart 1990). Both units have the ability to perform all the pipetting and heating steps required for a variety of sequencing applications and are user programmable. Both process 96 reactions at once in a microtitre well format starting from purified DNA. The Beckman unit has interchangeable pipetting heads and uses disposable tips, with a minimum pipetting volume of 2.0 µL. The Applied Biosystems unit has a single, fixed pipetting probe designed to achieve the greater pipetting accuracy required for the sub-microliter volumes used in microsequencing applications. Both units provide for hands-off running of standard sequencing chemistries.

Researchers at the Institute of Physical and Chemical Research in Japan have automated the sequencing process on a mega scale (Swinbanks 1991). This project was initiated by Akiyoshi Wada at the University of Tokyo in 1981and funded by the Japanese Science and Technology Agency and several private companies. The Human Genome Analyzer (HUGA) consists of a production line of machines that carry out the different steps of DNA analysis from DNA isolation to input of the final sequence data into a computer database. Computer-controlled robots transfer microtitreplates and gel casting cassettes between various stations on the analyzer, which can run unattended and generate up to 108,000 bp of sequence per day. This production-style sequencing is probably not suitable for the average lab's needs, but will provide a mechanism for large-scale sequencing such as the current Human Genome Project.

Southern Blotting

Southern blotting has traditionally been used to do restriction fragment length polymorphism (RFLP) analysis (Southern 1975; Sambrook *et al.* 1989). An illustration of the Southern blotting process is shown in Figure 2.3. First, highly purified DNA is digested with a restriction enzyme to form small pieces which are then subjected to agarose gel electrophoresis. Following electrophoresis, the gel is blotted onto a membrane so that the DNA fragments are transferred to a medium which is more easily handled than a fragile gel. These membranes are hybridized with ^{32}P-labeled

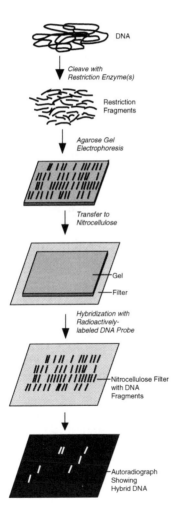

Figure 2.3
Schematic of the Southern blotting method.

probes to light up specific parts of the DNA, and the areas where the probes have hybridized are then detected by placing the membrane against a piece of X-ray film and performing autoradiography. This entire sequence of steps can take three or more days. Two common uses for Southern blotting in the clinical laboratory today are for assessing B- and T-cell rearrangements in malignant leukemia and lymphoma (Cossman *et al.* 1988) and in the molecular diagnosis of Fragile X mental retardation (Caskey *et al.* 1992).

There have been various approaches to automating different segments of this process (Winn-Deen 1992). Oncor (Gaithersburg, MD) has created an instrument line (Probe Tech™ 1 and Probe Tech™ 2) which facilitates the electrophoresis and blotting portion of the process. These instruments have a gel chamber and controls to set which reagents pass through that chamber. The chamber actually looks like a normal agarose gel electrophoresis chamber, except that the support beneath the gel is made of a porous plastic, allowing vacuum to be placed under the gel. Using this system, gels are poured in the usual way, and samples are loaded and electrophoresed. Following electrophoresis a piece of membrane is placed underneath the gel and the chamber is filled with depurination solution (dilute HCl). After a short time this solution is drained away by the Probe Tech and sodium hydroxide is added to denature the DNA. After drainage of this solution the chamber is filled with blotting buffer. Vacuum is applied through the porous gel support and DNA is drawn from the gel onto the membrane. Normally a transfer using capillary action is run overnight, but with vacuum transfer on this instrument all the DNA is completely transferred from the gel onto the membrane within 90 minutes, thus saving approximately 6 hours of lab time. The use of this chamber also minimizes the handling of the gel, which is important since broken gels make it much more difficult to get a good blot and to know exactly which band is where. The Probe Tech minimizes the handling of the gel and takes quite a bit of time off the entire blotting process. This system offers convenience for the first part of the Southern blotting process; however, the final step of hybridization to a labeled probe must still be done in the traditional manual manner. The Autotrans 7000 (Intelligenetics/ Betagen) and the TE 2000 Two Step™ (Hoefer) described in the section on DNA sequencing using radioactive labels electrophorese a sample down a vertical gel, and as the DNA elutes off the bottom of the gel, it is deposited onto a constantly moving membrane web. This approach is potentially simpler than the Oncor approach, and eliminates the need to handle the gel at all, but requires very precise tolerances on the moving parts. These are two different approaches to automation of the same part of the Southern blotting process.

Another part of the Southern blotting process that has been automated is the creation and analysis of the autoradiogram. As described previously, computing densitometers which can be used to measure the intensity of the bands from the exposed X-ray film are available from several manufacturers (Personal Densitometer™, Molecular Dynamics, Sunnyvale, CA; Bio Image™ Gel Scanner, Millipore, Bedford, MA; Optical Imaging System, Ambis, San Diego, CA). Other companies have automated the actual radiation measurement by replacing traditional autoradiography with direct reading of the radioactive blot (Betascope™, Intelligenetics/Betagen; Matrix™ , Packard Instruments, Meriden, CT; Ambis 1000/4000). If a radioactively labeled molecular weight standard has been included in a lane of the gel, the computer can also construct a standard curve, and then determine the size of any unknown bands. These instruments handle a variety of isotopes but have the drawback that samples which are weakly labeled may tie up the instrument for many hours counting a single blot. This problem is addressed by the phosphor imagers (PhosphorImager, Molecular Dynamics; Molecular Imager™, Bio-Rad, Hercules, CA). With these instruments each blot is placed in a cassette which contains a medium to capture the energy released by beta particles releasing photons. These photons are counted and their locations determined. These media have a greater dynamic range than conventional X-ray film and require much shorter exposure times. Since the instrument is used only for reading the medium, multiple blots can expose multiple phosphor plates simultaneously in a manner analogous to the autoradiographic part of the Southern blotting process. These systems are still a relatively expensive replacement for X-ray film.

Polymerase Chain Reaction (PCR)

The polymerase chain reaction (PCR) was the first of the target amplification procedures. A schematic of the method is shown in Figure 2.4. Two primers, one for each strand, are used to direct selective chain replication by DNA polymerase. As each chain is replicated it creates a copy of the orig-

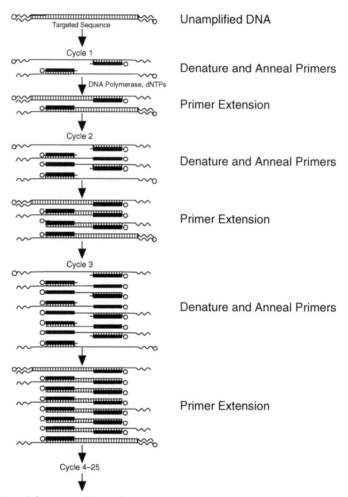

Figure 2.4. Schematic of the polymerase chain reaction amplification method (artwork courtesy of Perkin-Elmer Corporation).

inal template. Thermal denaturation of these copies from the original template results in four strands which can be replicated in the next cycle. By employing successive cycles of high temperature thermal denaturation and lower temperature primer annealing and chain extension, exponential amplification of the target sequence bracketed by the two primers is achieved. A thermal stable DNA polymerase permits this amplification procedure to be carried out in a programmable thermal cycler without the need to add more polymerase after each denaturation cycle (Lawyer *et al.* 1989).

This amplification method has been extensively developed since it was first described by Saiki *et al.* in 1985 (Bloch 1991; Erlich 1989; Elrich *et al.* 1989; Innis *et al.* 1990). PCR products are now used routinely for cloning and DNA sequencing. Diagnostic applications abound. The extreme sensitivity of this method (single copies of a target sequence can be detected) has led to its application in the detection of numerous infectious organisms, such as HIV, which have been difficult to culture. Applications in genetics include detection of point mutations with either competitive oligo priming (COP) (Gibbs *et al.* 1989) or amplification refractory mutation system (ARMS) (Newton *et al.* 1989) where mismatches between the primer and the mutant template result in loss of PCR product. Deletions and insertions are handled by bracketing the suspect area with PCR primers and observing the size of the PCR product formed. Translocation mutations, common in cancer, can be detected by choosing primers from the two different chromosomes involved in the translocation and observing whether any PCR product is formed. While the most common detection method is by gel electrophoresis, HPLC and microtitreplate capture methods have also been developed.

Synthesis of PCR Primers

PCR begins with synthesis and purification of the primers. Virtually all DNA synthesis today is done on an automated DNA synthesizer using phosphoramidite chemistry (Beaucage and Caruthers 1981; Matteucci and Carothers 1981; Vu *et al.* 1990). Automated synthesizers are available from a variety of manufacturers, including Applied Biosystems, Beckman Instruments (Fullerton, CA), Millipore (Bedford, MA) and Pharmacia. Solid supports containing the 3' base of the desired oligonucleotide (oligo) are successively derivitized 3' to 5' with the programmed sequence. At the end of the synthesis process the oligo, blocked at the 5' hydroxyl with a group such as dimethoxytrityl (DMT), is chemically cleaved from the support by treatment with concentrated ammonia. The cleaved oligo can now be deprotected with acetic acid either before or after purification. Purification is facilitated by leaving the DMT group in place, as only the correct, fully extended sequence will contain this group. All failure sequences are capped, typically by acetylation, before subsequent bases are added.

Analysis of oligo purity can be done by either high performance liquid chromatography (HPLC) (Eadie *et al.* 1987), or gel or capillary electrophoresis (CE) (Dubrow 1991). Both automated methods allow the user to load an autosampler and walk away while the oligo is sampled and separated. Each method uses a different physical property of the oligo to separate species. Trityl-on HPLC utilizes the difference in hydrophobicity between the tritylated oligo and nontritylated failure sequences. By varying the gradient and mobile phase conditions, a reverse phase column can separate the two species. If desired, preparative HPLC can be used to purify the desired oligo away from any contaminants. Separation by electrophoresis is based on the charge to mass ratio of the oligo, with shorter oligos eluting first. Single base resolution can be achieved, and with CE only a nanoliter of sample lost to the analysis. If electrophoresis is to be used for analysis of PCR products, CE separation will give an excellent preview of what one will expect to see on a standard slab gel.

Thermal Cycling

The PCR process requires a high temperature step for thermal denaturation followed by a lower temperature step to anneal and extend the primer. Originally this was accomplished by manually moving the PCR tube between heating blocks. This is not a particularly user-friendly way to run PCR, as a typical 30-cycle PCR protocol requires up to 90 block changes over a two to three hour period. The original approach to automation of thermal cycling, developed by Perkin-Elmer Corporation (Norwalk, CT), was a programmable heating block. Numerous other manufacturers have copied this approach. Perkin-Elmer has made some improvements on this method over the past few years, including engineering special thin-walled amplification tubes to facilitate thermal exchange between the block and the PCR reagent, and introducing a heated lid which eliminates the need to cover the PCR reaction with oil in order to prevent evaporation during thermal cycling. The capacity of these block style cyclers is generally 20–50 tubes or one 96-well microtitreplate. Most brands come with a built-in cooling source, although some lower priced brands require an external source. Other companies have taken slightly different approaches to automating thermal cycling. Stratagene (San Diego, CA) has recently introduced a robotic cycler which moves the tubes between blocks held at various temperatures in imitation of the original manual method. Others, such as BIOS (New Haven, CT) and Enprotech (Natick, MA), use forced air ovens to cycle. While this cycling method is considerably slower than the heating block approach since cooling is achieved passively, the capacity is greater and 4 microtitreplates or 200 tubes can be cycled simultaneously. Extremely rapid thermal cycling can now be done in capillary thermal cyclers such as those from Idaho Technology (Idaho Falls, ID) or A.B. Technology (Pullman, WA). By cycling in small thin-walled glass capillaries, cycle times of less than a minute are possible, greatly reducing total PCR time. While this is appealing in many ways, the process of loading and heat sealing the capillaries still requires some user-friendliness improvement. The decision on which thermal cycler to purchase should be made by carefully reviewing each laboratory's needs for amplification capacity and the sample-to-sample performance uniformity of the equipment under consideration.

Detection of PCR Products

High Performance Liquid Chromatography. High performance liquid chromatography (HPLC) has been used extensively to analyze and purify PCR primers. It can also be used to analyze and quantitate PCR product (Katz 1990; Katz and Dong 1990; Katz *et al.* 1993). Separation of a 500-mer product from primers and undesired side products, such as primer-dimers, can be done in about 20 minutes on an ion exchange column (for example, DEAE-NPR, Perkin-Elmer). Such rapid and quantitative analysis can be particularly useful during the optimization of a PCR assay, or for applications, such as RNA expression studies or assessment of viral load, where quantitation is required. Standard HPLC equipment already available in many labs (gradient maker, pump, injector, and UV detector) can be employed to analyze PCR reactions. By placing a guard column in-line with the separation column, PCR reactions from genomic DNA can be injected directly after amplification without any sample cleanup.

Gel Electrophoresis. PCR samples can be evaluated by simple gel elctrophoresis on agarose or acrylamide gels stained with an intercalating dye, such as ethidium bromide. They can also be quantitated by gel electrophoresis on a fluorescent DNA sequencer. Some sequencer manufacturers provide specialized software for fragment analysis (Genescan 672, Applied Biosystems; Fragment Analyzer™, Pharmacia). For this application PCR product can be labeled with either a fluorescently tagged primer or by staining the PCR product with ethidium bromide. Very small quantities of

DNA (100 attomoles) are required for analysis. For the Applied Biosystems sequencer, primers can be labeled in any of four dyes used for DNA sequencing. Specialized software provides both multicomponent analysis to separate the signals from the four colors and quantitation of the bands found during electrophoresis. The ability to analyze four colors simultaneously allows the inclusion of a dye-labeled molecular weight standard as an internal reference in each lane and automatic size calling of each band (Mayrand et al. 1992). This is especially important in identity testing where variations in electrophoretic mobility between lanes, or between gels run on different days or in different labs, can raise doubts whether an individual should be included or excluded as a result of the analysis. The Pharmacia system handles this same problem by allowing the computer to overlay bands from a molecular weight marker lane with a data lane, providing size calling of the fragments in the data lane.

In 1980 markers consisting of a variable number of tandem repeats (VNTRs) were first described by Wyman and White. The detection of these markers by PCR is achieved by placing the PCR primers outside the repeated region and amplifying across the region. The size of the resulting product is then related to the number of repeats within the region. These markers can be used to follow relatedness among family members, as well as for forensic analysis. Although the allele distribution is not even among the different repeat numbers, this system can still be used effectively to characterize individuals. Commonly used markers include the apolipoprotein B gene (Boerwinkle et al. 1989) with repeat units of 32 bases and a population distribution of 29–51 repeats, pYNZ22 (D17S30) on chromosome 17, a 70 bp repeat which produces PCR products ranging in size from 170–870 bp (Nakamura et al. 1988a; Horn et al. 1989), and pMCT118 (D1S80) on chromosome 1, a 16 bp repeat which produces PCR products ranging in size from 400–700 bp (Nakamura et al. 1988b). Smaller repeat units of 2–4 bases are also found sprinkled throughout the genome (Weber and May 1989; Edwards et al. 1991). By combining several of these independent markers a unique personal identification profile can be established. Multiplex PCR of VNTRs is emerging as a powerful tool in the forensics and paternity testing areas (Fregeau and Fournay 1993). Through use of multicolor fluorescent labeling of the PCR primers, these loci can be multiplexed and analyzed in a single lane of a gel. As the number of loci in the multiplex increases, so does the confidence that the fingerprint is indeed unique. Up to 15 of these VNTR loci have been multiplexed into a single lane of a fluorescent sequencer gel (5 VNTRs in each of 3 colors, all 3 colors in a single lane, with the fourth color used as a size standard in the same lane).

Quantitative fluorescent gel electrophoresis has also been successfully applied to the analysis of multiplex PCR reactions for Duchenne muscular dystrophy (DMD) (Kronick et al. 1990). The multicolor detection capability of the fluorescent DNA sequencer allows the 9 DMD PCR reactions (Chamberlain et al. 1988, 1990) labeled in blue to be run in the same lane with a VNTR marker (pYNZ22) for sample identification labeled in green and a molecular weight marker labeled in red (GENESCAN-1000 ROX, Applied Biosystems). This system provides verification that products of the predicted molecular weight have been made for all 9 DMD loci and an internal identity signature in the case of a later question of sample mix-up or complete deletion of the DMD gene. The increased sensitivity of this detection method allows a decrease in the number of PCR cycles from the 25 cycles required for visual interpretation of an ethidium bromide-stained gel. For multiplex PCR analysis of the DMD gene this is important as a decrease in the number of PCR cycles has been reported to minimize the amount of PCR artifacts, as well as lowering the problems encountered with maternal contamination of fetal samples (Chamberlain et al. 1988). Analysis of DMD families using this approach has also shown that automated fluorescent gel electrophoresis can be used for gene dosage studies of carriers as well as affected individuals who have deletions or duplications of specific portions of the DMD gene. Recently, gene dosage studies for aneuploidy analysis by automated fluorescent gel electrophoresis have also been reported (Mansfield 1993). This combination of size determination, quantitation, and internal standardization makes automated fluorescent gel electrophoresis a powerful tool for analysis of PCR products.

Another specialized area where automated gel electrophoresis has been used is in the detection of point mutations using single strand conformational polymorphism (SSCP) analysis of PCR products (Orita et al. 1989). In this technique the PCR product is denatured and allowed to assume a single stranded conformation which depends highly on the specific nucleic acid sequence. This conformation confers a specific electrophoretic mobility under non-denaturing conditions. Single base changes (mutations or polymorphisms) can cause visible shifts in this electrophoretic mobility. The Pharmacia PhastSystem™ allows the user to electrophorese the PCR products on a precast gel at different carefully controlled run temperatures. Typically two different temperatures, one cool (<10°C) and one nearer ambient, are run to maximize the chance of detecting conformational polymorphisms. The gel is then silver stained by the system (30–50°C, 45 min) and the bands visualized. Electrophoresis is rapid (<1 hour) and the sharp bands simplify SSCP interpretation. This technique has been applied successfully to point mutation detection in the *ras* and p53 oncogenes and phenylketonuria (Orita *et al.* 1989; Mohabeer 1991; Dockhorn-Dworniczak 1991).

Quantitation with Electrochemiluminescence. Sensitive post-PCR quantitation can also be done with electrochemiluminescence on the QPCR™ System 5000 (Perkin-Elmer, Norwalk, CT) (Anderson *et al.* 1993; DiCesare *et al.* 1993). For analytes which produce a single specific PCR product, the normal primers are replaced with one primer labeled with biotin and one primer labeled with tris (2,2'-bipyridine) ruthenium (II) (TBR) (Kenten *et al.* 1991, 1992). For analytes which produce both specific and nonspecific PCR products, one of the normal primers is replaced with a primer labeled with biotin and the PCR product is probed with a target-specific probe labeled with TBR. The TBR primers/probes are synthesized using a TBR phosphoramidite. After the PCR is completed, the biotinylated product is captured with streptavidin-coated magnetic beads. Products labeled with TBR during the PCR process are loaded directly into the QPCR™ System 5000 autosampler which can hold up to 50 samples for unattended operation. Samples which require probing are first denatured to generate single stranded DNA, hybridized with the TBR probe, and then captured with streptavidin-coated magnetic beads and placed into the autosampler. Once in the System 5000, the sample is aspirated into the detection cell and concentrated by the magnetic arm. QPCR™ assay buffer containing tripropylamine (TPA) flows through the cell to wash the beads. When the flow stops and voltage is applied, the TPA and TBR^{+2} are both oxidized to produce TPA^+ and TBR^{+3}. The TPA then deprotonates to yield a TPA^* free radical which in turn reacts with TBR^{+3} to produce an excited state TBR^{+2*}. The TBR^{+2*} decays to ground state, emitting light at 620 nm which is detected with a photomultiplier. After light measurement is complete, QPCR™ cell cleaner flows through the detection cell into a waste container and analysis of the next sample is begun. Analysis time in the System 5000 is about one minute per sample. This system has been applied to quantitation of HIV-1 by PCR (Kenten *et al.* 1992).

Conclusion

The molecular genetic methods described in this chapter have begun as research tools. Today most have also found their way into the clinical laboratory. Automation developed for research applications is also being used by clinical laboratories faced with the task of developing their own in-house assays. Over the next few years more and more assays and instrument systems will be commercialized specifically for the clinical laboratory. As time progresses, the tools for molecular genetic analysis will begin to resemble the more familiar clinical analyzers, and the actual assay technology will become more transparent to the user as the degree of automation increases. Eventually these new instruments will mate up with the laboratory information systems and will provide for direct down-loading of patient results into the central laboratory computer. While the exact timeframe is as yet unknown, one has only to look back at the evolution of the immunoassay from a radioactive, intensely manual procedure to the current rapid, automated, non-radioactive systems to realize that this evolution in inevitable.

References

Anderson, M.S., Di Cesare, J.L., Katz, E.D. 1993. An electrochemiluminescence-based detection system for quantitative PCR. *Am Biotech Lab* (July): 10

Ansorge, W., Sproat, B., Stegeman, J., Schwager, C., Zenke, M. 1987. Automated DNA sequencing: Ultrasensitive detection of fluorescent bands during electrophoresis. *Nucl Acids Res* 15: 4593-4602.

Beaucage, S.L., Caruthers, M.H. 1981. Deoxynucleoside phosphoramidites-a new class of key intermediates for deoxypolynucleotide synthesis. *Tet Letters* 22: 1859-1862.

Bloch, W. 1991. A biochemical perspective of the polymerase chain reaction. *Biochem* 30: 2735-2747.

Boerwinkle, E., Xiong, W., Fourest, E., Chan, L. 1989. Rapid typing of tandemly repeated hypervariable loci by the polymerase chain reaction: Application to the apolipoprotein B 3' hypervariable region. *Proc. Natl Acad Sci* USA 86: 212-216.

Cathcart, R. 1990. Advances in automating DNA sequencing. *Nature* 347: 310.

Caskey, C.T., Pizzuti, A., Fu, Y.-H., Fenwick, R.G. Nelson, D.L. 1992. Triplet repeat mutations in human disease. *Science* 256: 784-789.

Chamberlain. J.S., Gibbs, R.A., Rainer, J.E., Nguyen, P.N., Caskey, C.T. 1988. Deletion screening of the Duchenne muscular dystrophy locus via multiplex DNA amplification. *Nucl Acids Res* 16: 11141-11156.

Chamberlain, J.S., Gibbs, R.A., Rainer, J.E., Caskey, C.T. 1990. Multiplex PCR for the diagnosis of Duchenne muscular dystrophy. In *PCR Protocols: A Guide to Methods and Applications*, edited by Innis, M.A., Gelfand, D.H., Sninsky, J.J., White, T.J., San Diego, CA, Academic Press, pp. 272-281.

Civitello, A.B., Richards, S., Gibbs, R.A. 1992. A simple protocol for the automation of DNA cycle sequencing reactions and polymerase chain reactions. *DNA Sequence* 3: 178-23.

Connell, C., Fung, S., Heiner, C., *et al.* 1987. Automated DNA sequence analysis. *BioTechniques* 5: 342-348.

Cossman, J., Uppenkamp, M., Sundeen, J, Coupland, R., Raffeld, M. 1988. Molecular genetics and the diagnosis of lymphoma. *Arch Pathol Lab Med* 112: 117-127.

DiCesare, J., Grossman, B., Katz, E., Picozza, E., Ragussa, R., Woudenberg, T. 1993. A highly sensitive electrochemiluminescence-based detection system for automated PCR product quantitation. *BioTechniques* 15: 152-157.

Dockhorn-Dworniczak, B., Dworniczak, B., Brommelkamp, L., Bulles, J., Horst, J., Bocker, W.W. 1991. Non-isotopic detection of single strand conformation polymorphism (PCR-SSCP): a rapid and sensitive technique in diagnosis of phenylketonuria. *Nucl Acids Res* 19: 2500.

Dubrow, R.S. 1991. Analysis of synthetic oligonucleotide purity by capillary gel electrophoresis. *Amer Laboratory* (March): 64-67.

Eadie, J.S., McBride, L.J., Efcavitch, J.W., Hoff, L.B., Cathcart, R. 1987.

High performance liquid chromatography analysis of oligodeoxyribonucleotide base composition. *Anal Biochem* 165: 442-447.

Edwards, A., Civitello, A., Hammond, H.A., Caskey, C.T. 1991. DNA typing and genetic mapping with trimeric and tetrameric repeats. *Am J Hum Genet* 49: 746-756.

Erlich, H.A., ed. 1989, *PCR Technology: Principles and Applications for DNA Amplification.* New York, Stockton Press.

Erlich, H.A., Gibbs, R., Kazazian, H.H., eds. 1989, *Current Communications in Molecular Biology—The Polymerase Chain Reaction.* New York, Cold Spring Harbor Laboratory Press.

Fregeau, C.J., Fournay, R.M. 1993. DNA typing with fluorescently tagged short tandem repeats: A sensitive and accurate approach to human identification. *BioTechniques* 15: 100-119.

Gibbs, R.A., Nguyen, P.-N., McBride, L.J., Koepf, S.M., Caskey, C.T. 1989. Identification of mutations leading to the Lesch-Nyhan syndrome by automated direct DNA sequencing of *in vitro* amplified cDNA. *Proc Nat. Acad Sci* USA 86: 1919-1923.

Gibbs, R.A., Nguyen, P.-N., Caskey, C.T. 1989. Detection of single base differences by competitive oligonucleotide priming. *Nucl Acids Res* 17: 2437-2448.

Gibbs, R.A., Nguyen, P.-N., Edwards, A., Civitello, A.B., Caskey, C.T. 1990. Multiplex DNA deletion detection and exon sequencing of the hypoxanthine phosphoribosyltransferase gene in Lesch-Nyhan families. *Genomics* 7: 235-244.

Horn, G.T., Richards, B., Klinger, K.W. 1989. Amplification of a highly polymorphic VNTR segment by the polymerase chain reaction. *Nucl Acids Res* 17: 2140.

Innis, M.A., Gelfand, D.H., Sninsky, J.J., White, T.J., eds. 1990. *PCR Protocols: A Guide to Methods and Applications.* San Diego, Academic Press Inc.

Ishino, Y., Mineno, J., Inoue, T., *et al.* 1992. Practical applications in molecular biology of sensitive fluorescence detection by a laser-excited fluorescence image analyzer. *BioTechniques* 13: 936-943.

Johnston, R.F., Pickett, S.C., Barker, D.L. 1990. Autoradiography using storage phosphor technology. *Electrophoresis* 11: 355-360

Kaneoka, H., Lee, D.R., Hsu, K.-C., Sharp, G.C., Hoffman, R.W. 1991. Solid-phase direct DNA sequencing of allele-specific polymerase chain reaction-amplified HLA-DR genes. *BioTechniques* 10: 30-34.

Katz, E.D. 1990. Quantitation and purification of polymerase chain reaction products by liquid chromatography. *J Chrom* 512: 433-444.

Katz, E.D., Dong M.W. 1990. Rapid analysis and purification of products of polymerase chain reaction by high-performance liquid chromatography. *BioTechniques* 8: 546-555.

Katz, E.D., Haff, L.A., Eksteen, R. 1993. Rapid separation, quantitation and purification of products of polymerase chain reaction by high-performance liquid chromatography. In *Methods in Molecular Biology*, Vol. 15: *PCR Protocols: Current Methods and Applications*, edited by White, B.A., Totowa, NJ, Humana Press, pp. 63-74.

Kenten, J.H., Casadei, J., Link, J., *et al.* 1991. Rapid electrochemiluminescence assays of polymerase chain reaction products. *Clin Chem* 37: 1626-1632.

Kenten, J.H., Gudibande, S., Link, J., Willey, J., Curfman, B., Major, E.O., Massay, R. 1992. Improved electrochemiluminescent label for DNA probe assays: Rapid quantitative assays of HIV-1 polymerase chain reaction products. *Clin Chem* 38, 873-879.

Kronick, M., Ziegle, J., Robertson, J., Fenwick, R. 1990. Simultaneous detection of PCR products informative of human disease and human identity: A novel method for internal sample identification. *Am J Hum Genet* 47: A225.

Kovach, J.S., McGovern, R.M., Cassady, J.D., Swanson, S.K., Wold, L.E., Vogelstein, B., Sommer, S.S. 1991. Direct sequencing from touch preparations of human carcinomas: Analysis of p53 mutations in breast carcinomas. *J Natl Cancer Inst* 83: 1004-1009.

Lawyer, F.C., Stoffel, S., Saiki, R.K., Myambo, K.B., Drummond, R., Gelfand, D.H. 1989. Isolation, characterization, and expression in *Escherichia coli* of the DNA polymerase gene from Thermus aquaticus. *J Biol Chem* 264: 6427-6437.

Leren, T.P., Rodningen, O.K., Rosby, O., Solberg, K., Berg, K. 1993. Screening for point mutations by semi-automated DNA sequencing using Sequenase and magnetic beads. *BioTechniques* 14: 618-623.

Mansfield, E.S. 1993. Diagnosis of Down syndrome and other aneuploidies using quantitative polymerase chain reaction and small tandem repeat polymorphisms. *Hum Mol Genet* 2: 43-50.

Matteucci, M.D., Caruthers, M.H. 1981. Synthesis of deoxyoligonucleotides on a polymer support. *J Am Chem Soc* 103: 3185-3191.

Maxam, A.M., Gilbert, W. 1977. A new method for sequencing DNA. *Proc Nat Acad Sci USA* 74: 560-564

Mayrand, P.E., Hoff, L.B., McBride, et al. 1990. Automation of specific gene detection. *Clin Chem* 36: 2063-2071.

Mayrand, P.E., Corcoran, K.P., Zeigle, J.S., Robertson, J.M., Hoff, L.B. Kronick M.E. 1992. The use of fluorescence detection and internal lane standards to size PCR products automatically. *Applied and Theoretical Electrophoreis* 3: 1-11.

Middendorf, L.R., Bruce, J.C., Bruce, R.C., et al. 1992. Continuous, on-line DNA sequencing using a versatile infrared laser scanner/electrophoresis system. *Electrophoresis* 13: 487-494.

Mischiati, C., Fiorentino, D., Feriotto, G., Gambari, R. 1993. Use of an automated laboratory workstation for isolation of genomic DNA suitable for PCR and allele-specific hybridization. *BioTechniques* 15: 146-151.

Mohabeer, A.J., Hiti, A.L., Martin, W.J. 1991. Non-radioactive single strand conformation polymorphism (SSCP) using the Pharmacia Phast-System. *Nucl Acids Res* 19: 3154.

Nakamura, Y., Ballard, L., Leppert, M., et al. 1988. Isolation and mapping of a polymorphic DNA sequence (pYNZ22) on chromosome 17p. *Nucl Acids Res* 16: 5707.

Nakamura, Y., Carlson, M., Krapcho, V., White, R. 1988. Isolation and mapping of a polymorphic DNA sequence (pMCT118) on chromosome 1p (D1S80). *Nucl Acids Res* 16: 9364.

Newton, C.R., Graham, A., Hepinstall, L.E., et al. 1989. Analysis of any point mutation in DNA. The amplification refractory mutation system (ARMS). *Nucl Acids Res* 17: 2503-2516.

Olsvik, O., Wahlberg, J., Petterson, B., et al. 1993. Use of automated sequencing of polymerase chain reaction-generated amplicons to identify three types of cholera toxin subunit B in *Vibrio cholerae* O1 strains. *J Clin Micro* 31: 22-25.

Orita, M., Suzuki, Y., Sekiya, T., Hayashi, K. 1989. Rapid and sensitive detection of point mutations and DNA polymorphisms using the polymerase chain reaction. *Genomics* 5: 874-879.

Prober, J.M., Trainor, G.L., Dam, R.J., *et al.* 1987. A system for rapid DNA sequencing with fluorescent chain-terminating dideoxynucleotides. *Science* 238: 336-341

Saiki, R.K., Scharf, S., Faloona, F. *et al.* 1985. Enzymatic amplification of β-globin genomic sequences and restriction site analysis for diagnosis of sickle cell anemia. *Science* 230: 1350-1354.

Sambrook, J., Fritsch, E.F., Maniatis, T. 1989. *Molecular Cloning: A Laboratory Manual.* New York, Cold Spring Harbor Laboratory Press.

Sanger, F., Niklen, S., Coulsen, A.R. 1977. DNA sequencing with chain terminating inhibitors. *Proc Nat Acad Sci USA* 74: 5463-5467

Santamaria, P., Boyce-Jacino, M.T., Lindstrom, A.L., Barbosa, J.J., Faras, A.J., Rich S.S. 1992. HLA class II "typing": Direct sequencing of DRB, DQB, and DQA genes. *Hum Immunol* 33, 69-81.

Smith, L.M., Fung, S., Hunkapiller, M.W., Hunkapiller, T.J., Hood, L.E. 1985. The synthesis of oligonucleotides containing an aliphatic amino group at the 5' terminus: Synthesis of fluorescent DNA primers for use in DNA sequence analysis. *Nucl Acids Res* 13: 2399-2412.

Smith, L.M., Sanders, J.Z., Kaiser, R.J., *et al.* 1986. Fluorescence detection in automated DNA sequence analysis. *Nature* 321: 674-679

Southern, E.M. 1975. Detection of specific sequences among DNA fragments by gel electrophoresis. *J Mol Biol* 98: 503-517

Swinbanks, D. 1991. Japan's human genome project takes shape. *Nature* 351: 593

Weber, J.L., May, P.E. 1989. Abundant class of human DNA polymorphisms which can be typed using the polymerase chain reaction. *Am. J Hum Genet* 44: 388-396.

Wike, C.M., Korber, B.T.M., Daniels, M.R., *et al.* 1992. HIV-1 sequence variation between isolates from mother-infant transmission pairs. *AIDS Research and Human Retroviruses* 8: 1297-1300.

Wilson, R.K., Yuen, A.S., Clark, S.M., Spence, C. Arakelian, P., Hood, L.E. 1988. Automation of dideoxynucleotide DNA sequencing reactions using a robotic workstation. *BioTechniques* 6: 776-787.

Winn-Deen, E.S. 1992. Automation of electrophoretic techniques for DNA analysis-evolution of the Southern blot. *Lab Robotics and Automation* 4: 337-342.

Wolinsky, S.M., Wike, C.M., Korber, B.T.M., *et al.* 1992. Selective transmission of human immunodeficiency virus type-1 variants from mothers to infants. *Science* 255: 1134-1137.

Wyman, A.R., White, R. 1980. A highly polymorphic locus in human DNA. *Proc Natl Acad Sci USA* 77: 6754-6758.

Vu, H., McCollum, C., Jacobson, K., *et al.* 1990. Fast oligonucletide deprotection phosphoramidite chemistry for DNA synthesis. *Tet Letters* 31: 7269-7272.

CHAPTER 3

Molecular Techniques in Anatomic Pathology

Gregory J. Tsongalis, Ph.D.

Introduction

Anatomic pathology is the division of pathology that utilizes the principles of basic science to elucidate the etiology and pathogenesis of disease processes by thorough examination of tissues. At the gross level, the naked eye is used to examine whole tissues or organs. Microscopy, including all aspects of light and electron microscopy, is the next investigative step in correlating a disease process with abnormalities in cells or organelles. Pathologists are now faced with what could be viewed as the next frontier in diagnostic medicine, as extremely rapid advances in molecular biology provide the means to examine disease processes at the level of the nucleic acid. The impact which molecular biology will have on clinical laboratory medicine remains to be seen. There is an enormous amount of new molecular information being produced on what seems to be a daily basis and thus increasing numbers of genes as well as mutations available for examination by the clinical laboratory.

In this chapter, I will discuss those molecular techniques of value to the anatomic pathologist and their possible application to diagnostic laboratory medicine. One of these techniques will be a novel application of the polymerase chain reaction. My emphasis will be on molecular oncology and will provide an overview of the molecular pathology of three of the most common types of human cancers, breast, colon, and prostate cancers.

Molecular Oncology

It is well accepted that human malignancy develops as a multi-step process involving changes at the molecular level which activate cellular proto-oncogenes and inactivate tumor suppressor genes (Cline 1989; Bishop 1991; Croce 1991; Stubblefield 1991). Proto-oncogenes were originally described as normal genes with similar sequence to cancer-causing genes of retroviruses. This definition has been expanded to include genes not found in retroviruses but which are associated with neoplastic disease. Therefore, the term proto-oncogene refers to normal cellular genes involved in cellular proliferation and differentiation that may contribute to the development of cancer when the gene's sequence or expression is altered. Activation of these genes and subsequent stimulation of cell growth is now recognized as only one aspect in our understanding of carcinogenesis (Bishop 1991; Croce 1991). A second aspect of carcinogenesis is provided by a group of genes, tumor suppressor genes, which normally function to suppress abnormal or neoplastic growth (Hollingsworth and Lee 1991; Hunter 1991). Loss of function of these genes plays a significant role in development of malignant disease.

Numerous proto-oncogenes and tumor suppressor genes have been identified, while new genes which will determine one's susceptibility to various forms of neoplasia are still being discovered. A partial list of oncogenes and tumor suppressor genes is given in Table 3.1, as well as their gene functions and tumor involvement. The ability to detect mutations in these genes using a small amount of isolated DNA will allow the laboratory to determine patients' susceptibility or risk for developing specific neoplasias and be of significant assistance in determining prognosis as well as in monitoring therapies for existing disease. Lesions occurring at the gene level, including chromosomal translocations, gene amplifications or deletions, point mutations, repeat expansions and abnormal methylation patterns, are all detectable using the molecular techniques discussed below. After discussing the general applications of these molecular techniques, I will focus on the molecular pathology of breast, colon, and prostate cancers as specific examples of the types of information which can be obtained using these techniques.

Table 3.1. Representative genes associated with malignant disease.

Gene	Proposed Function	Neoplasms Involved
APC	Tumor suppressor	Familial adenomatous polyposis coli (FAP)
BCR/ABL	Protein tyrosine kinase	Chronic myelogenous leukemia (CML), acute lymphoblastic leukemia (ALL)
Brush-1	?	Breast cancer
BRCA-1	?	Breast cancer
DCC	Tumor suppressor	FAP, Acoustic neuroma, pheochromocytoma
ERBB-1	EGF receptor	Astrocytoma, sqamous cell carcinoma
ERBB-2 (NEU/ HER-2)	? Receptor	Renal cell carcinoma, adenocarcinoma of breast, ovary and stomach
FES	Protein tyrosine kinase	Ovarian cancer
FMS	Mutant CSF-1 receptor, protein tyrosin kinase	Hepatocellular carcinoma
FOS	Combines with JUN to form AP-1	Hepatocellular carcinoma
INT	Fibroblast growth factor	Adenocarcinoma of breast and prostate
JUN	DNA binding protein, AP-1	
MCC	Tumor suppressor	Colorectal carcinoma
hMSH2	DNA repair	Heredtary nonpolyposis colon cancer (HNPCC)
MYB	DNA binding protein	Adenocarcinoma of breast
MYC	DNA binding protein	Carcinoma of lung, breast, ovary, cervix, endometrium, prostate
NM23	?Metastasis regulator, nucleoside diphosphate kinase	Carcinoma of breast, colon
p53	Tumor suppressor	Carcinoma of breast, lung, ovary, colon
RAS (H-, K-,N-)	Membrane associated GTP binding, GTPase	Carcinoma of lung, cervix, endometrium, prostate, pancreas, thyroid, breast, liver
RB	Tumor suppressor	Retinoblastoma, Carcinoma of lung, breast, ovary
RET	Transmembrane tyrosine kinase	Multiple endocrine neoplasia, familial medullary thyroid carcinoma
WT-1	Tumor suppressor	Wilm's tumor, Rhabdomyosarcoma, carcinoma of breast

Isolation of Nucleic Acids

In 1953, Watson and Crick described the structure of DNA and its role as the informational "blueprint" for all living organisms. Numerous methods for extracting DNA from various sources including tissues, cells, and body fluids have been developed. DNA has also been extracted from paraffin-embedded tissues, making this vast resource of archival tissue available for retrospective molecular analysis. DNA is the most stable of the nucleic acids and thus more easily isolated without degradation of the sample occurring. This is partly because DNAses are generally heat sensitive and inactivated by incubation at 65°C for several minutes. RNAses, on the other hand, are extremely stable, difficult to inactivate, and therefore more likely to degrade RNA during the extraction process.

The evaluation of clinically important abnormalities at the molecular level begins with the isolation of DNA from either fresh or preserved human tissues (see chapter 2). The quality and quantity of DNA needed will be determined by the type of analysis to be done. Specific types of molecular analysis require the detection of large fragments of DNA, and therefore high molecular weight DNA is preferred as starting material (for Southern blotting). In cases where the DNA sequence of interest is of small size, then it is not as crucial to begin with high molecular weight DNA (for PCR). Within the nucleus of the cell, DNA is surrounded by both histone and non-histone proteins, some of which have high affinity for specific DNA sequences. These DNA/protein interactions must be abolished so that the DNA may be available as a substrate for various enzymes used in the molecular analysis. This requires a series of steps which include lysis of cells, protein dissolution with proteinases, organic or saturated salt extraction to remove peptides, and ethanol precipitation of the nucleic acids. The major difficulty in working with human genomic DNA is its tendency to be viscous in solution and its susceptibility to shearing forces. The latter is particularly true in formalin-fixed tissues where the protein/DNA interactions are more rigid. Extensive degradation of the DNA and thus target sequences is not compatible with molecular analysis.

Southern Blot Hybridization

In 1975, Southern described a method for detecting specific DNA sequences among DNA fragments which were separated by gel electrophoresis and hybridized with a labeled known DNA fragment of com-plementary sequence (see chapter 2). Genetic alterations occur in all tumor types and are often specific for a tumor type or related to the pro-gression of the cancer. Southern blot analysis has been used for the detec-tion of abnormalities, such as translocations, deletions, amplifications, and point mutations, in both oncogene and tumor suppressor gene sequences. Amplification or deletion of these specific genes in most types of neoplasias have been correlated with the clinicopathologic grade (loca-tion, metastases, degree of differentiation). For example, the myc oncogene is amplified and/or over-expressed in solid tumors, such as carcinoma of the breast, lung, cervix, colon, and urinary bladder (Bishop 1991). Over-expression of the myc gene is most often associated with a poor prognosis. Mutations in the p53 tumor suppressor gene, the most common genetic alterations in human cancers, have been detected by Southern blot analy-sis using allele-specific oligonucleotide probes (Hollstein et al. 1991; Yan-dell and Thor 1993). Increased copy number of c-erb-B2 (HER-2/Neu) in breast cancer has also been detected by Southern blot and is associated with larger tumor size, shorter relapse time, and lower survival rate (Poller and Ellis 1993).

The ras family of oncogenes (ha-ras, Ki-ras, N-ras) was the first to be iso-lated from human cancer. Point mutations in the ras genes produce a potentially transforming gene product and are found in most tumors. Mutated ras genes are tumorigenic when transfected into normal cells. Southern blot analysis for the Ha-ras gene in DNA isolated from several tumorigenic cell clones is shown in Figure 3.1. The single band in lanes 4 and 5 corresponds to the Ha-ras gene. The negative control is in lane 2 and 3 and the positive control (entire plasmid) is shown in lane 1.

Gene rearrangements (bcr/abl, T-cell receptor) and translocations (Philadelphia chromosome) in lympho-proliferative malignancies, such as chronic myelogenous leukemia (CML) and acute lymphoblastic leukemia (ALL), have been detected by Southern blot analysis (see chapter 2). The detection of DNA from an infectious pathogens can also be performed by Southern blot analysis.

Northern Blot Analysis

The Northern blot assay is the RNA counterpart of the Southern blot procedure, resulting in the detection of RNA sequences when the membrane is hybridized with a labeled probe. Although in principle these two procedures are very similar, there are distinct differences for reasons which include the single-strandedness of RNA, the fact that secondary structures of RNA must be denatured to ensure adequate electrophoretic mobility, and most importantly that RNAses are extremely stable enzymes. This last remark must be considered carefully as RNAses do not discriminate between easy or tedious and laborious RNA preparations.

Unlike the Southern blot, which provides evidence for the presence of a particular DNA sequence, the Northern blot can only be used to determine whether specific RNAs are present within the cells at the time of RNA isolation. mRNA constitutes approximately 1–5% of the cell's total RNA and is heterogenous in size and sequence, unlike rRNA, tRNA and snRNA. An increase in mRNA levels does not, however, necessarily indicate that there is increased transcription of a gene. Mutations within or associated with a gene sequence may result in alterations in the size of a transcribed mRNA or may abolish transcription of that gene altogether, thus leading to decreased expression. In other instances, mutations may result in increased amounts of mRNA produced from a particular gene and concomitantly increase the expression of this gene.

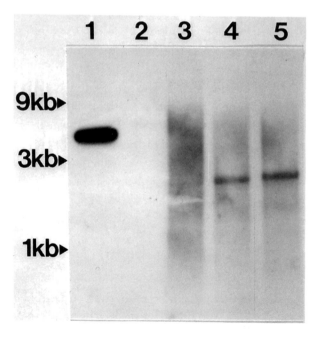

Figure 3.1. Southern blot analysis for the human H-ras gene. Lane 1, positive control; lane 2, negative control; lane 3, negative control; lanes 4 and 5, tumorigenic cell clones. (Courtesy of Dr. G.J. Smith and Dr. W.B. Coleman, Department of Pathology, University of North Carolina at Chapel Hill).

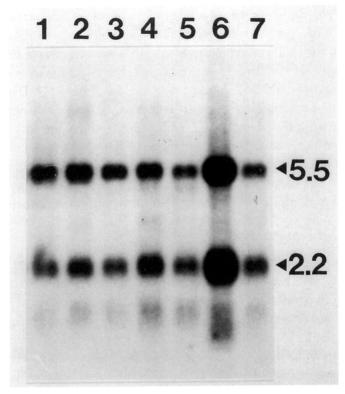

Figure 3.2. Northern blot analysis for the human K-ras mRNA. Lane 1, wild-type clone; lanes 2–7, transformed clones. Arrows indicate two K-ras mRNA species detected. (Courtesy of Dr. G.J. Smith and Dr. W.B. Coleman, Department of Pathology, University of North Carolina at Chapel Hill).

Mutations which activate oncogenes result in over-expression of the gene which has been shown to be associated with malignant disease and correlated with poor prognosis for specific types of cancers. Mutations of the Ki-ras gene have been found in 33% of adenocarcinomas of the lung, 90% of pancreatic adenocarcinomas, and 50% of colon adenocarcinomas. Northern blot analysis for Ki-ras gene expression is shown in Figure 3.2. Each lane contains mRNA isolated from different clones of an established cell line and was probed for the Ki-ras transcript. Lane 1 contains mRNA from the wild-type clone and exhibits the normal 2.2 kb and 5.5 kb Ki-ras bands. Lanes 2–7 contain mRNA from transformed cell clones of which only one (lane 6) exhibits over-expression of the Ki-ras gene. Thus, Northern blot analysis allows for the examination of gene expression and the correlation of this data with the morphologic and transformational characteristics at the cellular level. Elevated mRNA levels only suggest increased transcription of the gene. One must be careful in the interpretation of these types of results and should always include the mRNA levels of a gene (usually a housekeeping gene) not affected by the experimental conditions as an internal control. If mRNA levels of the internal control remain constant while the mRNA level of the gene of interest is increased, then this indicates increased transcription of the gene. However, a subsequent increase in translation and thus gene product is only suggested by increased transcription and must be confirmed at the protein level.

Slot/Dot Blot Analysis

Slot or dot blot analyses are essentially a modification of both the Southern and Northern blot analyses. DNA or RNA are bound directly to a solid support membrane and then hybridized with a specific probe. Unlike the Southern or Northern blot, in these procedures it is not a prerequisite to digest the DNA or RNA with restriction enzymes nor electrophoretically separate the nucleic acid fragments. Thus, the possibility of probing a large number of samples on a single membrane also offers a rapid method for screening purposes. The disadvantage, however, is that multiple size fragments which may hybridize to the probe are not resolved.

This technique affords one the opportunity to measure specific gene transcription as a function of cell state during time course studies. For example, mRNA may be isolated from cells or tissues after treatment with various experimental agents and the transcriptional response of the cell monitored over time.

Polymerase Chain Reaction

The polymerase chain reaction (PCR) has unequivocally been the most substantial molecular technique to evolve from the research laboratories since the discovery of restriction endonucleases and the Southern blot procedure. Crucial to the successes of PCR was the discovery of the Taq DNA polymerase, a thermostable enzyme of Thermus aquaticus which lives in the boiling geisers of Yellowstone National Park. PCR is an in vitro enzymatic method for the amplification of specific gene target sequences and is described in chapter 2.

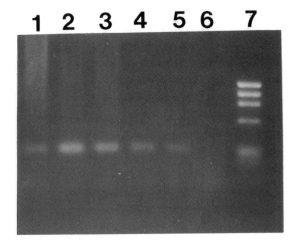

Figure 3.3. Agarose gel electrophoresis of PCR amplified exon 5 of the human p53 gene. Lanes 1–5, amplified product; lane 6, blank control; lane 7, molecular size markers.

A significant application of PCR is in the analysis of mutations in genomic DNA of eukaryotes. Amplification of a specific target sequence provides a substrate for numerous other analytical procedures, including restriction enzyme analysis and DNA sequence analysis. The polymerase chain reaction is primer-specific, with amplification of DNA sequences occurring between the forward and reverse primers. Nonspecific amplification can occur when primers anneal to non-complementary DNA sequences and these products can be avoided by altering several variables of PCR.

PCR is a technique by which one can rapidly detect and amplify a target sequence. This is of great advantage in cases where the presence of a specific target must be detrmined (e.g., infectious diseases). Amplification by PCR has made possible a wide variety of analyses ranging from mutation detection to DNA cloning. Agarose gel electrophoresis of amplified product from exon 5 of the human p53 gene is shown in Figure 3.3. Amplified PCR products are electrophoresed in agarose or acrylamide gels which are stained with ethidium bromide and visualized under ultraviolet light. The change in size of a PCR product created by a deletion or insertion can easily be demonstrated. Single base point mutations can be detected in amplifed PCR products, when followed by techniques such as denaturing gradient gel electrophoresis, single strand conformational polymorphism analysis, and chemical cleavage. These techniques have had greatest impact on prenatal diagnosis of genetic disorders and are currently being adapted to the detection of mutations in oncogenes and tumor suppressor genes for diagnostic purposes. Mutations in specific PCR-amplified DNA fragments may be identified by direct sequencing of the amplified product. Pathogenic organisms in clinical samples can also be identified by PCR technology as can biological samples for forensic purposes.

In Situ Hybridization

In situ hybridization (ISH) refers to the application of hybridization techniques to intact cells. Thus, a specific nucleic acid sequence may be localized to a cell within a tissue section or cytological preparation and correlated with the pathologic process.

The advantages to such a technique include cellular localization, specificity, sensitivity, elimination of processing fresh tissue to isolate nucleic acids, and quantitative ability. Nucleic acid probes have made ISH much more specific than the more traditional immunocytochemical methods which utilize monoclonal antibodies against structural elements of the cell. This technique becomes insensitive in cases of occult or latent viral infections or in cases of decreased mRNA levels where the number of possible target sequences per cell is low. In contrast, Southern blot hybridization at the expense of cellular localization may detect these low number of target sequences because isolated nucleic acid from the entire specimen is used in the assay.

This technique has been widely used and can be combined with fluorescent microscopy (FISH), electron microscopy, confocal microscopy, flow cytometry, or in situ amplification procedures which provide extremely powerful tools for examining tissues. ISH has been applied to various clinical diagnostic scenarios, including the detection of chromosomal aberrations, infectious organisms, single copy genes and mRNA. Human papillomavirus (HPV), Herpes simplex virus (HSV), human immunodeficiency virus (HIV), Cytomegalovirus (CMV), Epstein-Barr virus (EBV) and Adenovirus are among the more common viral organisms detected by ISH (Chang et al. 1993; Margall et al. 1993).

Approximately 65 human papillomaviruses have been identified, 23 of which infect the genital tract and are sexually transmitted. Specific HPV genotypes have been linked to characteristic pre-malignant and malignant anogenital lesions. Condylomata, squamous cell carcinoma and dysplastic lesions of the oral mucosa as well as laryngeal papillomata have been found to contain HPV DNA. ISH for HPV type 6/11 is shown in Fig-

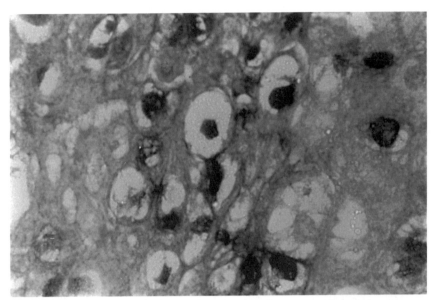

Figure 3.4. In situ hybridization for HPV types 6 and 11 in a tracheal papilloma. (Courtesy of Dr. K.U. Mehta, Department of Pathology, Monmouth Medical Center, Long Branch, NJ).

ure 3.4. HPV type 6/11 was detected in multiple confluent papillomas in the trachea of a patient with recurrent respiratory papillomatosis and was confirmed to be the etiologic agent by ISH (Mehta *et al.* 1991).

Localized In Situ Amplification (LISA)

The polymerase chain reaction provides the necessary technology to amplify specific DNA or mRNA sequences from a wide variety of sample sources. Until recently, the ability to correlate nucleic acid sequence identification with histological information has been accomplished by ISH. We have embarked on a technological journey to develop various strategies that allow us to closely associate amplified PCR product with the tissue of origin and more precisely with the pathology observed at the microscopic level.

In situ hybridization (ISH) provides a method by which specific target sequences may be identified within tissue sections. However, for the detection of single copy genes, latent viral infections, and various types of small mutations, the sensitivity of ISH is reduced because of lack of target sequence. If the target sequence were first amplified within the cell, then there would be ample target for ISH. An extremely powerful technique results from the combination of PCR and ISH. The coupling of these two techniques combines one method with extreme sensitivity and specificity (PCR) together with another method which provides the ability to determine cellular localization (ISH). Several investigators have described protocols for using these two techniques for the detection of viral genomic DNA as well as of single-copy genes in tissue samples and cytology specimens.

In situ amplification depends upon several crucial factors that are currently being addressed by ongoing research. Tissues which are to be examined using in situ amplification must be histologically prepared in such a way as to be optimal for PCR. For example, fixation and deparaffinization of embedded tissues must be conducted using protocols which will not inhibit the subsequent PCR reaction. Buffered formalin fixation appears to be near optimal for most applications and organosilane coated slides should be used to maintain attachment of the tissue during the procedure. Proteinase treatment of the sample is also required to make the DNA accessible to PCR reagents because formalin crosslinks proteins, forming a permeabilization barrier. The concentration and type of proteinase as well as the incubation time and temperature should be optimized for each application.

Once the tissue samples are prepared for PCR, several options become available. Amplification of DNA target sequences may be performed directly or, in the case of mRNA, a reverse transcription reaction may be performed prior to amplification of the cDNA. Subsequent to amplification the product must be identified at the microscopic level. A labelled nucleotide such as digoxigenin-dUTP may be added to the PCR reaction so that it becomes directly incorporated into the amplified product. Using immunohistochemical techniques, the product may then be identified and cellular localization visualized with light microscopy. Indirect methods such as ISH, where the target sequence is first amplified and then detected using probe technology, may also be used.

Initial attempts at "in situ" PCR followed by in situ hybridization were limited by the prerequisite of using cell suspensions in a tube amplification system (Haasse et al. 1990; Komminoth et al. 1992). Early in situ protocols were adopted for paraffin-embedded tissues, but these procedures had been problematic due to technical issues, such as destruction of morphology, optimization of preparative and amplification conditions for paraffin-embedded tissue sections, evaporation of reagents, and loss of localization of amplified product.

We have developed a method described as localized in situ amplification (LISA). This method allows direct amplification reactions to be performed, each localized to different regions of the same tissue section, by using tissue culture cloning rings as vessels for each reaction. This modification to in situ amplification reactions has helped circumvent many of the previously encountered technical difficulties. LISA should prove very useful for both research and diagnostic applications.

LISA can be performed on a variety of tissue samples, including paraffin-embedded tissues, cytospin cell preparations, tissue touch preparations, and cytological specimens, thus serving a large array of interests. We have found that the morphology of specimens remains of good quality. Depending upon which thermal cycler is used, uneven heat distribution may need to be accounted for. We perform LISA using a thermal cycler (COY TempCycler II) which contains a specifically designed thermal block proportional in size to a standard histological glass slide, so that an even distribution of heat is maintained throughout the slide (Figure 3.5).

By using tissue culture cloning rings, it is possible to localize the amplification reaction to the area of interest on the tissue section or to one of several tissue sections mounted on the same slide (Figure 3.6). Each ring provides a separate chamber for amplification, allowing several amplification reactions on the same tissue section to be performed. Thus, several control reactions can be performed simultaneously. The use of these rings to form an external chamber which is placed on top of the tissue section simulates to some degree a tube amplification reaction and thus allows one to optimize amplification conditions on genomic DNA in a more traditional fashion.

A major concern regarding in situ molecular procedures is the evaporation of reagents and the destruction of the tissue with manipulation of the coverslip. Although clear nail polish sealing the coverslip has been used successfully to decrease evaporation of the reaction mixture, it has made manipulation of the coverslip difficult. In LISA the cloning ring is easily positioned onto the tissue section and requires minimal nail polish to form a seal between the ring and glass slide (Figure 3.6). Removal of the ring is

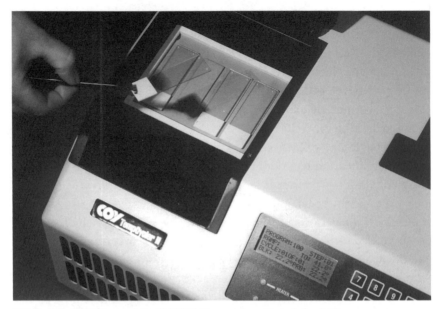

Figure 3.5. COY Tempcycler II (COY Corporation, Grass Lake, MI).

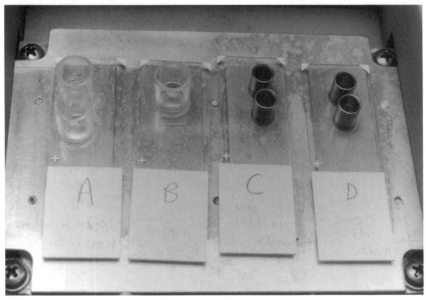

Figure 3.6. Localized *in situ* amplification (LISA) using the COY Tempcycler II.

not required for the remainder of the amplification procedure. It is removed during the detection process so that tissue outside the ring can function as a control for nonspecific binding of antibody or chromagen. Evaporation of reagents is eliminated by layering mineral oil or an Ampliwax bead over the reaction as in routine PCR.

Non-specific amplification is a possibility in any type of amplification protocol used. We have found that this is not the case when detecting the directly incorporated labeled nucleotide. However, difficulty arises from detection of labeled nucleotide which has non-specifically *incorporated* into cellular genomic DNA at DNA gaps and breaks, for example. Therefore, extreme care must be taken to ensure that the proper control reactions have been performed simultaneously and the results interpreted cautiously. In any in situ amplification reaction, one should include both positive and negative controls. The omission of Taq polymerase from a reaction is not sufficient as a negative control. This is shown in Figure 3.7 (see color figure insert following page 56). When LISA was performed on both lung and liver sections using primers for a segment of the human CFTR gene, nuclear signal of amplified product was observed (Figures 3.7A and 3.7C). If Taq polymerase was omitted from the reaction, no nuclear signal was seen (Figures 3.7B and 3.7D). This indicates that the target sequence is amplified when both polymerase and primers are present. If the polymerase is omitted from the reaction as a negative control, then no product is detectable. However, due to non-specific incorporation of the labeled nucleotide, these same results are obtained in the absence of primer but presence of polymerase. Therefore, the omission of specific primers or the inclusion of a set of primers with no known target sequence in the sample are much more appropriate and significant as negative controls.

It is also possible to omit other components of the amplification reaction for negative control reactions. An example is shown in Figure 3.8A (see color figures following page 56). Normal colon tissue touch preparations provided a sample source for a LISA reaction using alpha-1-antitrypsin gene primers. Note the prominent nuclear staining. When magnesium was omitted from the amplification reaction very little background nuclear staining was detected (Figure 3.8C). This then suggests that the product detected in Figure 3.8A is indeed amplified target sequence. Similar results to those in Figure 3.8C were obtained if the primers were omitted as the negative control reaction. No non-specific staining is observed in these samples outside the amplification ring (Figure 3.8B), indicating that the results in Figure 3.8A are not due to aberrent antibody or chromagen binding.

Positive controls should include a tissue section known to contain the target sequence. For example, LISA was peformed on rat lung tissue from a known positive case of Pneumocystis carinii pneumonia using primers specific for the rRNA gene of P. carinii. Clusters of P. carinii organisms within alveolar spaces were identified after immunohistochemical staining for digoxigenin-dUTP incorporated into amplified product (Figure 3.9A) (see color figures following page 56). When primers were omitted from the amplification reaction, no detectable product was observed (Figure 3.9B). Hematoxylin/eosin staining of the tissue section showed foamy clusters of P. carinii within the alveolar spaces. The presence of P. carinii was also confirmed by the more traditional silver stain method (Figure 3.9C). P. carinii provides an excellent model system for developing LISA because it is an extracellular organism; thus confusion of amplified product with background nuclear staining is avoided. Also, there are special histochemical stains for this organism so that the presence of target sequence can be confirmed.

As described above, our modifications of in situ amplification procedures described by others have established LISA as a more user-friendly technique that avoids many of the technical problems associated with in situ amplification. Under proper conditions, the coupling of in situ amplification with in situ detection protocols provides a sensitive method for detecting abnormalities at the molecular level and correlating them with the pathologic process.

Direct Mutation Detection Techniques

Mutations, which may consist of as little as a single base change, can be detected by a variety of chemical, physical, and/or enzymatic means. Today, most of these techniques require amplification of a target sequence by PCR, followed by scanning techniques to detect unknown mutations or techniques which allow identification and quantification of known mutations.

Denaturing gradient gel electrophoresis (DGGE) uses the abnormal denaturation behavior of mismatches in heteroduplex complementary DNA strands derived from wild type and mutant sequences as the basis for determining the presence of a mutation (Landegren 1992). Chemical

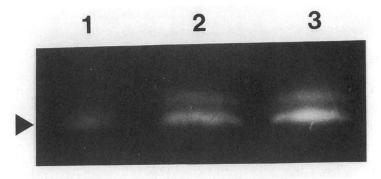

Figure 3.10. Heteroduplex analysis of p53 exon 7 in a Li-Fraumeni Syndrome patient using MDE hydrolink gel electrophoresis. Arrowhead indicates the wild-type allele.

modification and cleavage relies on chemical alteration of mismatched bases between the wild type and mutant sequence which render them susceptible to cleavage at the site of the mismatch. The effect of mutations on the migration characteristics of a specific DNA fragment can be detected by the single strand conformation analysis (SSCA) or by heteroduplex analysis. Altered migration in SSCA relies on the altered secondary structure of single-stranded DNA due to a single base change while that in heteroduplex analysis relies on conformational differences in double-stranded DNA due to heteroduplex formation between the mutant and wild type strands (Prior *et al.* 1993).

DGGE analysis and SSCA have been used routinely by laboratories to detect the presence of point mutations in the p53 tumor suppressor gene with subsequent sequence analysis to identify the mutation. We have opted to detect the presence of p53 mutations using heteroduplex analysis because of the simplicity of the technique. p53 is the most commonly mutated gene in human cancers, and we wanted to assess the ability of heteroduplex formation to detect the presence of known p53 mutations. Exon 7 of the human p53 gene was amplified from DNA isolated from a patient with Li-Fraumeni syndrome known to have a point mutation at codon 248. The amplified product was denatured and slowly reannealed to enhance heteroduplex formation before electrophoresis in an MDE hydrolink gel (ATGC Biochem, Malvern, PA). This is shown in Figure 3.10, where the arrowhead indicates the wild type allele and the heteroduplex band can be easily seen in the patient samples (lanes 2 and 3). This technology offers a rapid method for mutation screening in genes associated with disease and may be suitable for large population screening studies. DNA sequence analysis is required, however, to determine the type of mutation.

Molecular Pathology of Solid Tumors

Molecular characterization of solid tumors will provide a great challenge to the molecular biologist because neoplastic disease has proven to be a multi-step process involving numerous genes. In addition, most tumors comprise a heterogeneous population of cells, despite the clonal nature of neoplasms due to invasion of surrounding normal tissue.

Breast Cancer

Carcinoma of the breast remains the major incident cancer in females, accounting for approximately 32% of all female cancer incidences in 1993. The etiology of breast cancer is extremely complex and involves numerous genetic, endocrine, and environmental factors. Although the majority of breast cancers occur from acquired mutations, 5% of breast cancers result from inherited mutations. Molecular information concerning these types of neoplasias will thus play a role in diagnostic and therapeutic evaluations, as well as in determining if a patient has a predisposed increased risk of developing breast cancer.

Analysis of primary breast tumor tissue has revealed that genetic alteration of proto-oncogenes resulted in the amplification of the genes int2, myc, and erbB-2/HER2/neu (Dickson and Slamon 1990). Over-expression of c-erbB-2 has been associated with resistance to chemotherapeutic drugs and thus poor prognosis (Poller and Ellis 1993). Nm23, a cancer metastasis gene, has been shown to have reduced expression in the more aggressive breast cancers (Steeg et al. 1993). Mutations in the p53 tumor suppressor gene have been reported in 50% of all breast cancers and loss of heterozygosity (LOH) at this locus occurs in 50% of primary breast tumors (Osborne et al. 1991). LOH of chromosome 1, 3, 11, 17, and 18 are common in human brest cancers. A predisposing tumor suppressor gene for both breast and ovarian cancers, BRCA1, has been mapped to the long arm of chromosome 17 (17q12-21) (Hall et al. 1990; Smith et al. 1992; Black et al. 1993; Feunteun et al. 1993; Saito et al. 1993). This region has been shown to account for the largest proportion of inherited breast cancers.

Colorectal Cancer

Colorectal cancers have provided an excellent system in which to search for molecular alterations, because they have a progressive history and are associated with several inherited syndromes. Unlike other common human tumors, highly malignant tumors and more benign precursor lesions are available for study from the same patient with colorectal cancer. The colonic mucosa gives rise to small adenomatous polyps of low malignant potential which may then give rise to larger adenomas with increased risk of malignant transformation. Ultimately, these adenomas result in malignant neoplasms with invasive and metastatic potential. It is now well accepted that the development of colorectal cancers involves the accumulation of multiple genetic lesions in both oncogenes and tumor suppressor genes. Elucidation of these multi-step pathways are currently being pursued for sporadic and inherited cancers via methodologies, including some of those discussed earlier in this chapter.

Familial adenomatous polyposis (FAP) is an inherited autosomal dominant syndrome characterized by the development of hundreds to thousands of adenomatous polyps in the colon and rectum. A small fraction of these adenomas then progress to advanced colon cancer. Linkage studies identified a chromosomal band, 5q21, which was deleted in FAP patients. This led to the identification of a possible new tumor suppressor gene, APC, which was mutated in the germline of these patients (Groden *et al.* 1991; Joslyn *et al.* 1991). Localized mutations and large deletions have been identified in the APC gene that inactivate the gene through creation of frameshifts, stop codons, or missense mutations. In colorectal cancers from patients with no known familial predisposition, the 5q21 allele is deleted in 35–60% of cases. A second gene, MCC (**mutated in colorectal cancer**) is also located at 5q21 and has been implicated in the development of sporadic colorectal cancers (Kinzler *et al.* 1991). Other genetic alterations include ras mutations in 50% of colon cancers as well as in 50% of adenomas greater than 1 cm in diameter. LOH of 18q21 in 70% of tumors has led to the identification of the DCC gene, another tumor suppressor gene involved in the development of colon cancer (Fearon *et al.* 1990). Most recently, the hMSH2 gene has been identified at the 2p22-21 locus. This gene has been associated with hereditary nonpolyposis colon cancer (Fishel *et al.* 1993; Peltomaki *et al.* 1993).

Prostate Cancer

According to the American Cancer Society prostate cancer will account for approximately 28% of new male cancer incidences in 1993. More frequently encountered than the lethal neoplasms of the prostate are those forms of prostate cancer discovered as incidental findings at postmortem examinations or in surgical pathology specimens. Prostatic neoplasia and hyperplasia have been traditionally evaluated by light microscopic techniques that reveal characteristic morphological features of prostate tumors. Histologically, these neoplasms are characterized by histologic-grade heterogeneity, admixtures of benign and malignant glands, varying amounts of surrounding stromal tissue, and multifocality.

Molecular analyses of prostate cancers have demonstrated the possible pathogenic role of one or more tumor suppressor genes by frequent loss of heterozygosity (LOH) of chromosome regions 8p, 10q, 13q, 16q, 17p, and 18q. LOH of chromosome regions 10q and 16q have been found in 30% of all adenocarcinomas of the prostate (Wolman *et al.* 1992). The p53 tumor suppressor gene is often mutated in these tumors. Alterations in other oncogenes resulting in their activation has been well documented in prostate cancer. These include point mutations in the ras gene family and and have also resulted in elevated sis, myc, and fos mRNA levels. The int-2 gene, whch codes for the basic fibroblast growth factor, has also been shown to have elevated expression in these neoplasms.

Conclusion

Without doubt, molecular biology will have a great impact on laboratory medicine and diagnosis of human disease. In this chapter I have only highlighted some of the applications of molecular techniques in the diagnosis and prognosis of human cancers, mainly breast, colon and prostate cancers. As we begin to unravel the molecular mechanisms of human disease processes, we will be able to approach the challenge of correlating phenotype with genotype in the hopes of providing better diagnostic testing and therapies.

References

Bishop, J.M. 1991. Molecular themes in oncogenesis. *Cell* 64:235-248.

Black, D.M., Nicolai, H., Borrow, J., Solomon, E. 1993. A somatic cell hybrid map of the long arm of human chromosome 17, containing the familial breast cancer locus (BRCA1). Am J Hum Genet 52:702-710.

Chang, F., Syrjanen, S., Shen, Q., Wang, L., and Syrjanen-K. 1993. Screening for human paillomavirus infections in esophageal squamous cell carcinomas by in situ hybridization. Cancer. 72 (9): 2525-30.

Cline, M.J. 1989. Molecular diagnosis of human cancer. Lab Invest 61:368-380.

Croce, C.M. 1991. Genetic approaches to the study of the molecular basis of human cancer. Cancer Res 51:5015s-5018s.

Dickson, R.B., Slamon, D. 1990. New insights into breast cancer: The molecular biochemical and cellular biology of breast cancer. Cancer Res 50:4446-4447.

Fearon, E.R., Cho, K.R., Nigro, J.M., et al. 1990. Identification of a chromosome 18q gene that is altered in colorectal cancers. Science 247:49-56.

Feunteun, J., Narod, S.A., Lynch, H.T., et al. 1993. A breast-ovarian cancer susceptibility gene maps to chromosome 17q21. Am J Hum Genet 52:736-742.

Fishel, R., Lescoe, M.K., Rao, M.R.S., et al. 1993. The human mutator gene homolog MSH2 and its association with hereditary nonpolyposis colon cancer. Cell 75:1027-38.

Groden, J., Thilveris, A., Samowitz, W., et al. 1991. Identification and characterization of the familial adenomatous polyposis coli gene. Cell 66:589-600.

Haasse, A.T., Retzel, E.F., Staskus, K.A. 1990. Amplification and detection of lentiviral DNA inside cells. Proc Natl Acad Sci USA 87:4971-4975.

Hall J.M., Lee M.K., Morrow J., Anderson L., King M.C. 1990. Linkage of early onset familial breast cancer to chromosome 17q21. Science 250:1684-1689.

Hollingsworth, R.E., Lee W.H. 1991. Tumor suppressor genes: New prospects for cancer research. J Natl Cancer Inst 83:91-96.

Hollstein, M., Sidransky, D., Vogelstein B., Harris, C. 1991. p53 mutations in human cancers. Science 253:49-53.

Hunter, T. 1991. Cooperation between oncogenes. Cell 64:249-270.

Joslyn, G., Carlson, M., Thilveris, A., et al. 1991. Identification of deletion mutations and three new genes at the familial polyposis locus. Cell 66:601-613.

Kinzler, K.W., Nilbert, M.C., Vogelstein, B., et al. 1991. Identification of a gene located at chromosome 5q21 that is mutated in colorectal cancers. Science 251:1366-1370.

Komminoth, P., Long, A.A., Ray, R., Wolfe, H.J. 1992. In situ polymerase chain reaction detection of viral DNA, single-copy genes, and gene rearrangements in cell suspensions and cytospins. Diag Mol Pathol 1:85-97.

Landgren, U. 1992. Detection of mutations in human DNA. Genet Anal Tech Appl 9:3-8.

Margall, N., Matias-Guiu, X., Chillon, M., et al. 1993. Detection of human papillomavirus 16 and 18 DNA in epithelial lesions of the lower genital tract by in situ hybridization and polymerase chain reaction: Cervical scrapes are not substitutes for biopsies. J Clin Microbiol 31:924-930.

Mehta, K.U., Bolanowski, P.J.P., Combates, N., Raska, K. 1991. Tracheal papillomatosis: Molecular diagnostic techniques and human papillomavirus. NJ Med 86:187-189.

Osborne, R.J., Merlo, G.R., Mitsudomi, T., et al. 1991. Mutations in the p53 gene in primary human breast cancers. Cancer Res 51:6194-6198.

Peltomaki, P., Aaltonen, L.A., Sistonen, P., et al. 1993. Genetic mapping of a locus predisposing to human colorectal cancer. Science 260:810-12.

Poller, D.N., Ellis, I.O. 1993. Oncogenes and tumor morphology prediction. Mod Pathol 6:376-377.

Prior, T.W., Papp, A.C., Snyder, P.J., et al. 1993. Identification of two point mutations and a one base deletion in exon 19 of the dystrophin gene by heteroduplex formation. Hum Mol Genet 2:311-313.

Saito, H., Inazawa, J., Saito, S., et al. 1993. Detailed deletion mapping of chromosome 17q in ovarian and breast cancers: 2-cM region on 17q21.3 often and commonly deleted in tumors. Cancer Res 53:3382-3385.

Smith, S.A., Easton, D.F., Evans, D.G.R., Ponder, B.A.J. 1992. Allele losses in the region 17q12-21 in familial breast and ovarian cancer involve the wild-type chromosome. Nature Genet 2:128-131.

Steeg, P.S., De La Rosa, A., Flatow, U., MacDonald, N.J., Benedict, M., Leone, A. 1993. Nm23 and breast cancer metastasis. *Breast Cancer Res Treat* 25:175-187.

Stubblefield, E. 1991. The genetic changes in cancer. *Mol Carcinog* 4:257-260.

Wolman, S.R., Macoska, J.A., Micale, M.A., Sakr, W.A. 1992. An approach to definition of genetic alterations in prostate cancer. *Diag Mol Pathol* 1:192-199.

Yandell, D.W., Thor, A.D. 1993. p53 analysis in diagnostic pathology. *Diag Mol Pathol* 2:1-3.

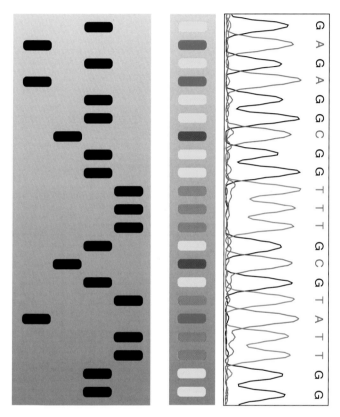

Figure 2.2. Schematic comparing four-lane sequencing using radioactive labels with one-lane sequencing using four fluorescent labels (artwork courtesy of Applied Biosystems). See page 12.

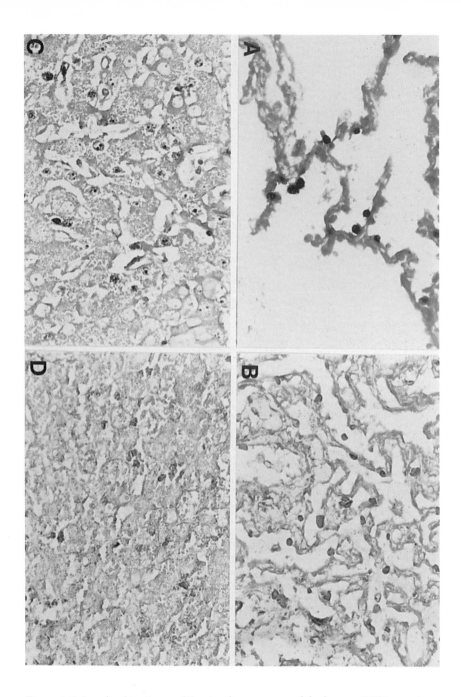

Figure 3.7. Localized in situ amplification for a segment of the human CFTR gene in lung (A and B) and liver (C and D). Panel A and C, with primers and Taq polymerase; panel B and D, with primers but without Taq polymerase. See page 47.

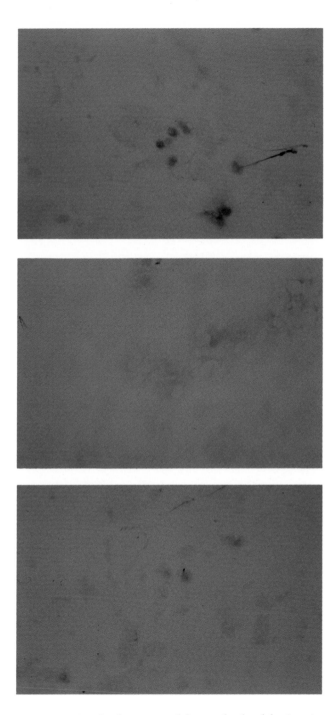

Figure 3.8. Localized in situ amplification for the alpha-1-antitrypsin gene in normal colon touch preparations, 8A (top). Tissue outside the cloning ring, 8B (middle); amplification without magnesium, 8C (bottom). See page 47.

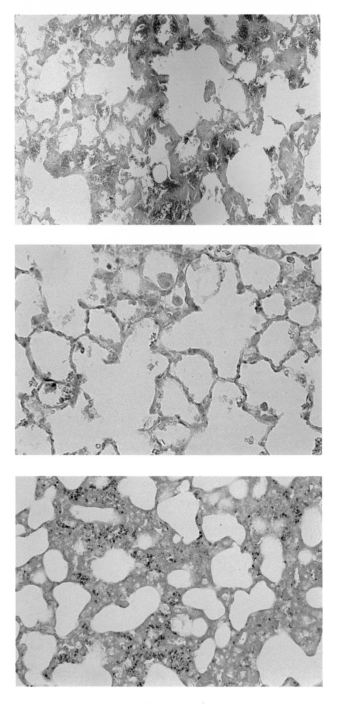

Figure 3.9. Localized in situ amplification of the rRNA sequence for Pneumocystis carinii in rat lung tissue, 9A (top). Without primers, 9B (middle); confirmation by silver staining, 9C (bottom).

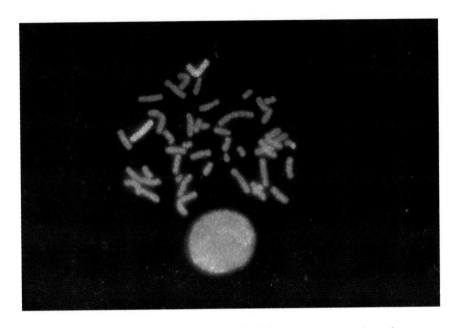

Figure 5.4. Fluorescence *in situ* hybridization (FISH) of a human metaphase chromosome spread, using a chromosome 3 "paint" probe. The patient has an unbalanced translocation, with a karyotype of 46XX, -22, +der(22)t(3;22)(p24.3;p11). (Photograph courtesy of Dr. Kathleen Rao, University of North Carolina, Chapel Hill). See page 91.

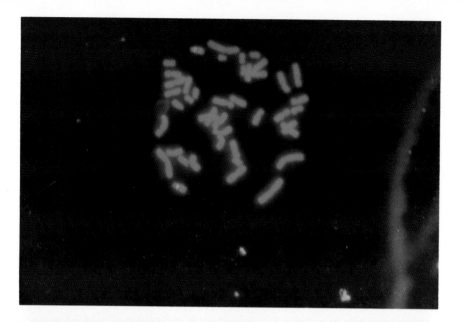

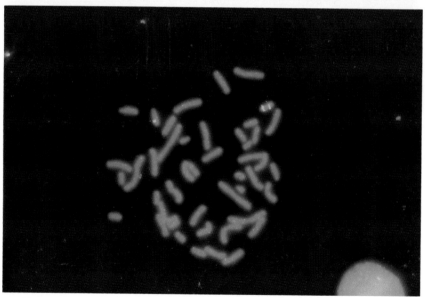

Figure 5.5. Fluorescence *in situ* hybridization (FISH) of human metaphase chromosome spreads, using a unique sequence probe specific for the Prader-Wili/Angelman region on chromosome 15. (a) normal karyotype. (b) patient with a submicroscopic deletion of the Prader-Wili/Angelman region and a karyotype of del(15)(q11q13). (Photographs courtesy of Dr. Kathleen Rao, University of North Carolina, Chapel Hill). See page 91.

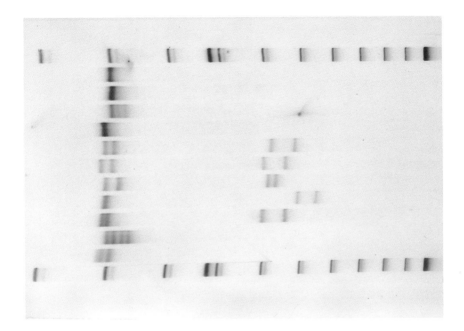

Figure 6.4. Detection of CFTR exon 10 mutations and polymorphisms by MDE™ gel electrophoresis. See page 106.
Exon 10 of the CFTR gene was amplified from patients with known genotypes using the PCR primers and conditions of Zielinski *et al.* (59). $^1/_{10}$th of tht reaction mixture was electrophoresed on a 40 cm x 1 mm MDE™ (AT Biochem, Malvern, PA) gel for 20,500 Volt-hours. Under these conditions, the bottom bands (homoduplexes, 491 bp) migrate approximately 37 cm into the gel. The amplified products were detected by silver staining. Reprinted from *Clinical Chemistry*: 1993; 39; Figure 1: 706-7, courtesy of the American Association for Clinical Chemisty, Inc.

Lane #	Patient Genotype	Lane #	Patient Genotype
1.	BRL 100 bp Ladder	7.	ΔI507/Wild Type
2.	M470V Homozygote Allele 1	8.	ΔF508/I506V
3.	M470V Homozygote	9.	ΔF508/F508C
4.	M470V Heterozygote	10.	ΔF508/Q493X
5.	ΔF508/ΔF508	11.	I506V/Wild Type
6.	ΔF508/Wild Type	12.	F508C/Wild Type
		13.	BRL 100 bp Ladder

CHAPTER 4

Molecular Pathology and Hematopoietic Neoplasia

Georgette A. Dent, M.D.

Introduction

Discoveries using the techniques of molecular biology have made an enormous contribution toward elucidating the pathogenesis of hematopoietic neoplasms. The techniques are also playing an increasingly important role in the diagnosis and monitoring of hematopoietic diseases. In 1992, the College of American Pathologists began a pilot program to gather information on the use of Southern blotting in diagnostic hematology laboratories. This is an indication of the increasing acceptance of this technology in the clinical laboratory.

Molecular biological techniques have four primary applications in the diagnostic hematology laboratory. These are:

1. detection of clonal populations
2. determination of cell lineage
3. detection of translocations
4. detection of minimal residual disease.

The two techniques generally utilized to address these applications are Southern blotting and the polymerase chain reaction (PCR).

We first introduced Southern blotting into the Clinical Hematology Laboratory at the University of North Carolina Hospitals in 1987. Since then we have performed over 300 diagnostic procedures. We currently offer four Southern blot-based tests as a diagnostic service: detection of rearrangement of the immunoglobulin heavy chain and T-cell receptor β chain genes, detection of the *bcr/abl* translocation, and detection of the *bcl-2* translocation. DNA probes to detect the first three molecular alterations currently enjoy the widest usage in clinical hematology laboratories, are FDA approved, and are readily available from commercial sources. At the UNC Hospitals, we also offer detection of the *bcl-2* and *bcr/abl* translocations using the PCR for investigational purposes. In this chapter we will review the use of molecular probes to T- and B-cell gene rearrangements and translocations specific for hematopoietic neoplasms, and discuss their clinical applications.

Immunoglobulin Gene Rearrangements and B-cell Neoplasia

The biology of the process by which the immunoglobulin and T-cell receptor genes rearrange has been the subject of several excellent reviews (Alt, Blackwell and Yancopoulos 1987; Griesser *et al.* 1989; Sklar 1992) and will be only briefly reviewed here. Immunoglobulin molecules are heterodimers consisting of two heavy chains and two light chains, and each of the four chains has a variable and a constant region. The variable region is the part of the molecule involved in antigen recognition and binding. The immunoglobulin heavy chain gene resides on chromosome 14, and in its germline configuration consists of several non-identical variable (V_H), diversity (D_H), and joining (J_H) regions. As part of B-cell differentiation, the three immunoglobulin genes (the heavy chain gene, and the κ and λ light chain genes) undergo the process of rearrangement. The rearrangements are accomplished with the help of a lymphocyte-specific recombinase, which is an enzyme that is able to recognize conserved nucleotide sequences that flank the DNA segments capable of rearrangement. Only one VDJ_H unit is needed to code for the variable portion of the immunoglobulin gene, and in the process of rearrangement, the extraneous DNA regions are lost. The ability of the immunoglobulin gene to generate different combinations of the V_H, D_H, and J_H regions, and therefore different variable regions, is one of the mechanisms by which antibody diversity is generated. The immunoglobulin gene rearrangements are virtually unique to an individual B-cell and are replicated in all cellular progeny. For this reason, the presence of a rearranged band is evidence of a clonal population. Normal polyclonal B-cell populations will have so many rearrangements that only the unrearranged germline band will be detected using Southern blotting technology.

The genetic structure of the κ and λ light chains is similar to that of the heavy chain gene. The κ light chain gene resides on chromosome 2 and the λ light chain gene resides on chromosome 22. In the germline state, each of these genes consists of several V and J regions. Unlike the immunoglobulin heavy chain gene, no D region is present in the light chain genes. After rearrangement of the immunoglobulin heavy chain gene, the maturing B lymphocyte attempts to rearrange one of the four light chain genes, beginning first with the two κ chain gene loci. This process terminates with the first functional rearrangement in a process known as allelic exclusion. If rearrangement of the κ chain gene is unsuccessful, the B lymphocyte will attempt to rearrange the lambda light chain. Approximately 60% of the time, rearrangement of the κ light chain gene is successful. This accounts for the predominance of κ–producing B-cells in normal B-cell populations as well as among B-cell neoplasms.

Rearrangements of the immunoglobulin genes are found in the entire spectrum of B-cell malignancies, from immature precursor B-cell lymphoblastic leukemias to mature B-cell lymphomas and leukemias (reviewed in Medeiros, Bagg and Cossman 1992). Virtually all B-cell neoplasms will have a rearranged immunoglobulin heavy chain gene, and mature B-cell neoplasms will have both the heavy chain gene and one of the light chain genes rearranged. Because of the developmental hierarchy in which the immunoglobulin genes rearrange, it is unusual for a B-cell neoplasm to have light chain rearrangement in the absence of heavy chain gene rearrangement (Williams *et al.* 1987).

Southern blotting technology, for the detection of gene rearrangements, is an important component of the diagnostic work-up of a lymphoid neoplasm. Southern blots have the ability to detect a clonal population that consists of 1–10% of the total cellular population and are therefore more sensitive than the human eye in detecting small populations (Aisenberg 1993). In our laboratory, the diagnostic work-up of a lymphoma may include flow cytometry, frozen section immunoperoxidase studies, paraffin section immunoperoxidase studies, cytogenetics and molecular diagnostic testing, using Southern blotting in addition to routine histology. Lymph node biopsies and extranodal tissues are received in the laboratory in the fresh state. The lymph node is examined macroscopically and a sample is obtained for flow cytometry. Touch imprints are made, tissue is submitted for routine histology, and a section is frozen for frozen section immunoperoxidase studies and/or molecular diagnostic testing. Our experience agrees with that of others who have found that in most cases, the diagnosis of a non-Hodgkin's lymphoma can be established with a combination of morphologic and immunophenotypic studies (Henni et al. 1988). Although the diagnosis of a lymphoid neoplasm can generally be established with a combination of morphology and immunophenotyping, in certain situations, gene rearrangement studies may be invaluable. For example, gene rearrangement studies are particularly helpful in the diagnosis of extranodal lymphocytic proliferations (Knowles et al. 1989).

One area in which the utility of molecular studies is controversial is in the diagnosis of Hodgkin's disease. A variety of molecular abnormalities have been reported in Hodgkin's disease, including immunoglobulin and T-cell receptor gene rearrangements (Griesser et al. 1989) and the bcl-2 translocation (Gupta et al. 1992). A recent study using single cell PCR analysis has elegantly demonstrated the heterogeneous nature of the genetic alterations that can be associated with Hodgkin's disease (Trumper et al. 1993).

At the UNC Hospitals, we use the J_H probe (Oncor) to detect monoclonal B-cell populations. This is a probe to the constant region of the immunoglobulin heavy chain gene and it has FDA approval. A standard Southern blot is prepared, in which the DNA is extracted from tissue samples, peripheral blood or bone marrow using a non-organic DNA extraction method (Oncor); the DNA is quantitated and digested with restriction endonucleases following the manufacturer's instructions. The digested DNA is loaded onto a 0.7% agarose gel, the samples are electrophoresed, and transfer is to a nylon membrane using the Probe Tech I from Oncor. A photograph of the gel is examined to ensure that the DNA has been digested and is not degraded. While the gel is being electrophoresed and transferred to a nylon membrane, the J_H probe is radiolabelled with ^{32}P. The membranes are then hybridized with the radiolabelled probe. The membranes are washed and autoradiographed with an intensifying screen for 1–7 days depending on the specific activity of the ^{32}P.

The gels are always interpreted in conjunction with clinical, morphologic, and immunophenotypic data. We require a rearranged band to be present in two of three restriction digests to call the procedure positive for the

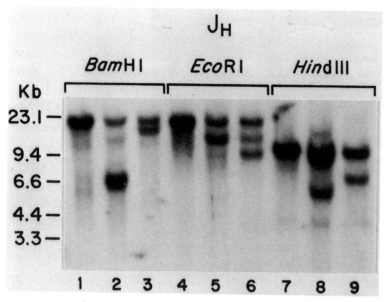

Figure 4.1. Southern blot analysis for immunoglobulin heavy chain gene rearrangement using a probe to the constant region of the gene (J_H). The DNA samples have been digested with three restriction enzymes as indicated. The probe identifies bands of 18 kb, 18 kb, and 11 kb when hybridized to germline genomic DNA digested with EcoRI, BamHI, and HindIII respectively. DNA samples from placenta (lanes 1, 4, and 7) show the expected germline pattern seen in DNA used as negative controls in the procedure. Rearranged bands can be seen in lanes 2, 3, 5, 6, 8, and 9, indicating monoclonal B-cell populations in DNA isolated from non-Hodgkin's lymphoma tissue samples.

presence of a monoclonal population. Alternatively, the presence of two rearranged bands with one single enzyme is also indicative of a positive result (Cossman *et al.* 1991). Figure 4.1 is an example of an autoradiograph of a Southern blot that was hybridized with the J_H probe to the immunoglobulin heavy chain gene. The J_H probe identifies bands of 18 kb, 18 kb, and 11 kb when hybridized to germline genomic DNA digested with EcoRI, BamHI, and HindIII respectively. DNA samples from placenta (lanes 1, 4, and 7) have been digested with BamHI, EcoRI, and HindIII as indicated. Lanes 1, 4, and 7 represent the expected pattern seen in DNA used as negative controls in the procedure. Rearranged bands can be seen in lanes 2, 3, 5, 6, 8, and 9, indicating monoclonal B-cell populations in DNA isolated from non-Hodgkin's lymphoma tissue samples. Issues to be concerned with in interpreting this procedure include cross-hybridization bands and polymorphisms (Cossman *et al.* 1991; Farkas 1993). For this reason different restriction enzymes must be used and the appropriate positive, negative and sensitivity controls should be included (Cossman *et al.* 1991; Farkas 1993). Added sensitivity and specificity can be obtained by using probes to the light chain genes; however, the majority of B- cell neoplasms can be detected using probes to the heavy chain gene. It is extremely unusual for the heavy chain gene to be rearranged without rearrangement of a light chain gene although this phenomenon has been reported (Cleary, Warnke and Sklar 1984).

The use of Southern blotting in a clinical setting suffers from several drawbacks. The procedure requires a highly trained staff, is expensive, and takes one to two weeks if the procedure goes well. Another drawback is that Southern blotting requires relatively large amounts of high molecular weight DNA. Because of the disadvantages inherent in Southern blotting, several groups have utilized the PCR to detect rearrangements of the immunoglobulin and T-cell receptor genes (Kuppers *et al.* 1993; Segal *et al.* 1992; Trainor *et al.* 1991). PCR methodology offers several advantages over Southern blotting. It is more rapid, does not generally require radioactivity, is more sensitive and requires less DNA. For these reasons, Southern blotting for the detection of T- and B-cell gene rearrangements will in all likelihood be replaced by PCR technology.

T-cell Receptor Gene Rearrangements and T-cell Neoplasia

Just as detection of rearrangements of the immunoglobulin heavy chain genes can be used to detect monoclonal populations of B-cells, so can detection of rearrangements of the T-cell receptor genes be used to determine monoclonal populations of T-cells. The organization of the T-cell receptor genes is analogous to that of the immunoglobulin genes. There are two T-cell receptors that each exist as heterodimers: the $\alpha\beta$ receptor and the $\gamma\delta$ receptor (reviewed in Strominger 1989). Approximately 98–99% of mature T-cells, of both helper and suppresser phenotype, express the $\alpha\beta$ receptor and the remaining 1–2% express the $\gamma\delta$ receptor. The organization of the four T-cell receptor genes ($\alpha,\beta,\gamma,\delta$) is similar to that of the immunoglobulin genes. T-lymphocytes undergo an analogous process of gene rearrangement during differentiation. Each of the four T-cell receptor genes is composed of several V and J regions in their germline configura-

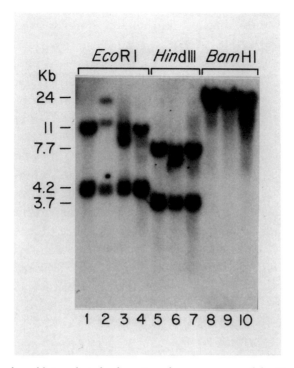

Figure 4.2 Southern blot analysis for detection of rearrangement of the T-cell receptor beta chain gene using a probe to the constant region of the gene ($C_T\beta$). The DNA samples have been digested with three restriction enzymes as indicated. The probe identifies bands of 11 kb and 4.2 kb, 7.7 kb and 3.7 kb, and 24 kb when hybridized to germline genomic DNA digested with EcoRI, HindIII, and BamHI respectively. Lanes 1, 4, 5, 7, 8, 9 and 10 show the germline pattern. Lane 10 also shows a cross hybridization band. Rearranged bands are seen in lanes 2, 3, 6, and 9. Lanes 3, 6 and 9 contained DNA isolated from a lymph node biopsy obtained from a patient with a Ki-1 (CD30)-positive anaplastic large cell lymphoma. The rearranged bands seen with the EcoRI and HindIII restrictions are indicative of the T-cell lineage of this lymphoma.

tions, and the T-cell receptor β and δ chain gene contains several D regions as well. During T-lymphocyte differentiation, the T-cell receptor δ gene rearranges first, followed by the γ, β, and α genes. The most widely used probes for detection of rearrangement of the T-cell receptor genes are probes to the T-cell receptor β chain gene. Because there is no marker of clonality for T-cells as there is for B-cells, this technology is particularly useful for the diagnosis of T-cell neoplasms (Knowles 1989). This technology can be utilized in differentiating angioimmunoblastic-like T-cell lymphomas from angioimmunoblastic lymphadenopathy (Frizzera 1992) and in establishing the diagnosis of the large granular lymphocyte (LGL) leukemia (Loughran Jr. 1993).

The procedure we follow at the UNC Hospitals is similar to that outlined above. We use a probe to the β chain of the T-cell receptor. In addition to cross-hybridization bands and polymorphisms, incomplete restrictions are particularly problematic with this probe (Cossman et al. 1991; Farkas 1993).

Figure 4.2 shows an autoradiograph of a Southern blot hybridized with a probe to the T-cell receptor beta chain gene, the $C_T\beta$ probe (Oncor). This probe identifies bands of 11 kb and 4.2 kb, 7.7 kb and 3.7 kb, and 24 kb when hybridized to germline genomic DNA digested with EcoRI, HindIII, and BamHI respectively. Lanes 1, 4, 5, 7, 8, 9, and 10 show the germline pattern. Lane 10 also shows a 17.5 kb cross-hybridization band. Lanes 3, 6, and 9 contained DNA isolated from a lymph node biopsy obtained from a patient with a Ki-1 (CD30)-positive anaplastic large cell lymphoma. The rearranged bands seen with the EcoRI and HindIII restrictions are indicative of the T-cell lineage of this lymphoma.

Translocations and B-cell Lymphomas

Other tests useful in the diagnosis and classification of lymphomas include detection of the bcl-2 translocation or t(14;18), the bcl-1 translocation or t(11;14), and the translocations found in Burkitt's lymphoma involving the c-myc oncogene and one of the immunoglobulin genes (t(2;8), t(8;14), and t(8;22)).

An excellent review of the bcl-2 gene and its protein can be found in Korsmeyer (1993). The t(14;18) is a balanced translocation involving chromosomes 14 and 18 and is found in approximately 90% of follicular lymphomas and in approximately 30% of diffuse large cell lymphomas (Weiss et al. 1987). The translocation is generally restricted to two regions referred to as the major break point or mbr (Tsujimoto et al. 1984) and the minor cluster region or mcr (Cleary, Gallili and Sklar 1986). As a result of the juxtaposition of chromosomes 14 and 18, the bcl-2 gene, which normally resides on chromosome 18, is placed under the transcriptional regulation of the immunoglobulin locus. The bcl-2 protein resides in the mitochondrial inner membrane and it functions to block programmed cell death or apoptosis. The translocation results in deregulation of the 25 kilodalton bcl-2 protein. B-cell lymphomas with the bcl-2 rearrangement

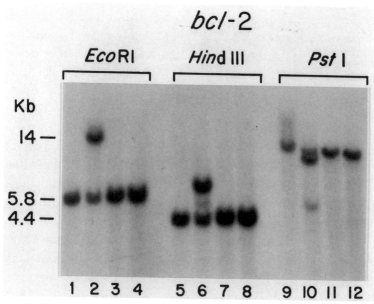

Figure 4.3. Southern blot analysis for the t(14;18) translocation using a DNA probe to the mbr region on chromosome 18. The DNA has been digested with three restriction enzymes as indicated. This probe identifies bands of 5.8 kb, 4.4 kb and 14 kb respectively when hybridized to germline genomic DNA digested with EcoRI, HindIII, and PstI respectively. Lanes 1, 3, 4, 5, 7, 8, 9, 11, and 12 show the germline (unrearranged) pattern. Lanes 2, 6, and 10 show rearranged bands indicative of the t(14;18) translocation in a patient with a follicular lymphoma.

express increased amounts of the bcl-2 protein. The bcl-2 translocation is virtually unique to B-cell neoplasms of follicular center cell origin and can serve as a marker of clonality and B-cell lineage. Although the detection of this translocation may play a role in the classification and staging of B-cell lymphomas (Gulley, Dent and Ross 1992), there is apparently no relationship between the presence of this translocation and prognosis in follicular lymphomas (Pezella et al. 1992). Southern blotting technology using probes to translocations involving c-myc, bcl-1 or bcl-2 can diagnose approximately 35% of the lymphomas that occur in North America (Asinberg 1993).

Figure 4.3 shows an autoradiograph of a Southern blot hybridized with a probe that recognizes the mbr region on chromosome 18, the bcl-2 probe. This probe detects both the t(14;18) and the germline bcl-2 gene on chromosome 18 and identifies bands of 5.8 kb, 4.4 kb and 14 kb respectively when hybridized to germline genomic DNA digested with *Eco*RI, *Hind*III, and *Pst*I respectively. Lanes 1, 3, 4, 5, 7, 8, 9, 11 and 12 show the germline (unrearranged) pattern. Lanes 2, 6, and 10 show rearranged bands associated with the t(14;18) rearrangement in a patient with a follicular lymphoma.

Because the *bcl-2* translocation is generally restricted to two limited areas of chromosome 18, it can be detected using PCR. However, since the *bcl-2* translocation has been found in benign lymphoid hyperplasia, detection of this translocation using sensitive PCR methodologies should be interpreted with caution when attempting to differentiate benign lesions from malignant processes (Limpens *et al.* 1991). One area in which the detection of the *bcl-2* translocation by PCR should be extremely useful is in detecting minimal residual disease in patients with non-Hodgkin's lymphoma who are candidates for autologous bone marrow transplantation. A major concern during autologous transplantation is that tumor cells will be reinfused into the patient. Patients with follicular lymphoma who have been in remission for years following conventional chemotherapy may remain PCR positive for the *bcl-2* translocation even though their marrows may have no morphologic evidence of lymphoma (Price *et al.* 1991). Gribben and co-workers have shown that patients who are PCR-positive for the *bcl-2* translocation after bone marrow purging are at an increased risk for relapse of their non-Hodgkin's lymphoma after autologous bone marrow transplantation (1991), and they have demonstrated the utility of using PCR to compare the effectiveness of different methodologies for purging the bone marrow of lymphoma cells prior to autologous bone marrow transplantation (1992). A recent study conducted by their group suggests that disease-free survival is increased in patients whose bone marrow samples are PCR-negative for the *bcl-2* translocation after autologous bone marrow transplantation, but that a subgroup of patients who are PCR-positive for the *bcl-2* early after transplantation may not be at an increased risk for relapse (Gribben *et al.* 1993). Additional studies should clarify the role of minimal residual disease detection in determining the appropriate management of patients undergoing bone marrow transplantation.

PCR of the *bcl-2* translocation is performed using a primer to either the mbr or mcr regions with a primer to the immunoglobulin heavy chain gene (JH) consensus region. For added sensitivity and specificity it is best to utilize a specific detection system which generally involves transferring

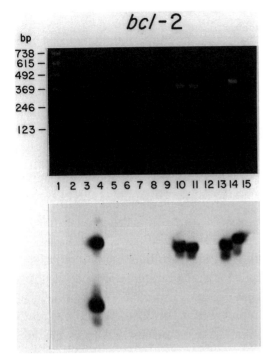

Figure 4.4. Southern blot analysis for *bcl-2* using PCR products amplified with primers to the mbr region and the immunoglobulin JH consensus region. The upper panel shows an ultraviolet light (UV)-illuminated ethidium bromide stained 2.0% agarose gel. The lower panel shows an autoradiograph from a Southern transfer of this gel after hybridization with an internal mbr oligonucleotide probe. A size marker is seen in lane 1. Lanes 2, 4-9, 11 and 12 are negative for the *bcl-2* translocation. Lane 3 shows the results using DNA isolated from the SU-DHL-4 cell line which contains the t(14;18) translocation. Lanes 10, 11, 14 and 15 show positive results from patients with follicular lymphomas.

the PCR product to a membrane and hybridizing it with a specific internal probe (Segal *et al.* 1994). Figure 4.4 shows an ultraviolet light (UV)-illuminated ethidium bromide-stained 2.0% agarose gel containing mbr/JH PCR products and an autoradiograph from the Southern analysis of this gel. The probe and primer sequences are shown in Table 4.1 (Benitez *et al.* 1992; Cleary, Smith and Sklar 1986). The PCR reaction products were electrophoresed, transferred to a nylon membrane, hybridized with a digoxigenin-labeled internal mbr oligonucleotide probe, and detected using a non-radioactive chemiluminescence detection system according to the manufacturer's instructions (Boehringer Mannheim). Lane 3 shows the results of probing DNA isolated from the SU-DHL-4 cell line which contains the t(14;18) translocation (Cleary, Smith and Sklar 1986). Lanes 10, 11, 14 and 15 show positive results from patients with follicular lymphomas. Lanes 2, 4-9, 11 and 12 are negative for the bcl-2 translocation. A size marker is seen in lane 1 (BRL).

Table 4.1. Oligonucleotides used for bcl-2 PCR

Name	Sequence
mbr primer[a]	5'-GGCCTATACACTATTTGTGACC-3'
JH primer[b]	5'-CACCTGAGGAGACGGTGACC-3'
mbr probe[a]	5'-GGTGATCGTTTTCTGTTTGAGA-3'

a from Cleary et al. 1986.
b from Benetiz et al. 1992.

The bcr/abl Translocation and Hematopoietic Neoplasia

The bcr/abl translocation associated with the Philadelphia chromosome (Ph[1]) and chronic myelogenous leukemia (CML) (Nowell and Hungerford 1960) is easily detected by Southern blotting and the PCR. The bcr/abl juxtaposition or the t(9;22) (q34;q11) is a reciprocal translocation resulting in the relocation of the c-abl gene on chromosome 22 to a location adjacent to the bcr gene on chromosome 9 (Rowley 1973). The Ph[1] can result from either one of two distinct t(9;22) translocations that differ on the molecular level. The t(9;22) associated with CML results in a bcr/abl fusion gene and chimeric bcr/abl mRNA that is translated into a 210 kilodalton protein (p210) (Gale and Canaani 1984). The presence of this translocation and protein is characteristic of CML. Approximately 3–5% of childhood ALL, 25% of adult ALL and 2–3% of adult de novo AML will also have a t(9;22) (reviewed in Kurzock, Gutterman and Talpaz 1988). The t(9;22) has also been reported in a rare case of non-Hodgkin's lymphoma (Cheng et al. 1992). Of the t(9;22) translocations associated with acute lymphoblastic leukemia (ALL), approximately half are identical to the translocation associated with CML and half are the result of a translocation 5' to the one typical of CML, resulting in a 190 kilodalton protein (Chan et al. 1987). Both chimeric proteins are associated with an abnormal tyrosine kinase activity, as compared with the normal 145 kilodalton c-abl protein, and are thought to play a pivotal role in the pathogenesis of hematopoietic neoplasia (reviewed in Heisterkamp and Groffen 1991).

We use a DNA probe known as the TransProbe-1™ (Oncogene Science) or phl/bcr-3 to detect the t(9;22) associated with CML (Blennerhassett et al. 1988). Figure 4.5 shows an autoradiograph of a Southern blot hybridized with the TransProbe-1™. This probe identifies bands of 4.8 kb, 3.2 kb,

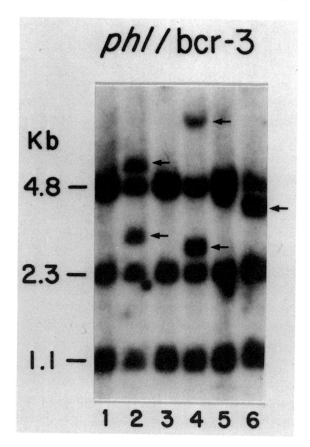

Figure 4.5. Southern blot analysis for the *bcr/abl* translocation using the *phl*/bcr-3 probe. This probe identifies bands of 4.8 kb, 3.2 kb, and 1.1 kb when hybridized to germline DNA digested with *Bgl*II. Lanes 2, 4, and 6 show rearranged bands in patients with CML. The rearranged bands are indicated by the arrows.

and 1.1 kb when hybridized to germline DNA digested with *Bgl*II. Lanes 2, 4, and 6 show rearranged bands in patients with CML.

One advantage of using Southern blots to detect the t(9;22) is the capacity of Southern blots to detect complex translocations (Ayscue *et al.* 1990). Complex translocations involve juxtapositions of the *bcr* and *c-abl* genes that are not easily detected using conventional cytogenetic analysis. It is now generally accepted that the presence of the *bcr/abl* translocation and the p210 protein are pathognomonic for CML, and disorders without molecular evidence of the *bcr/abl* juxtaposition most likely represent other myelodysplastic or myeloproliferative disorders (Pugh *et al.* 1988; Travis, Pierre and DeWald 1986). Because of the specificity of this translocation for CML, we use this test to differentiate CML from other forms of hematopoietic neoplasia and from leukemoid reactions.

In addition to being detectable using Southern blotting technology, the *bcr/abl* translocation can also be detected using reverse transcription (RT)-PCR (Lee *et al.* 1988). This is done by using reverse transcriptase to transcribe the chimeric mRNA into cDNA and then performing the PCR on the cDNA. As in the case with *bcl-2* and non-Hodgkin's lymphoma, this is potentially a powerful technique for monitoring minimal residual disease in patients who are candidates for bone marrow transplantation (Dhringa *et al.* 1992).

Figure 4.6 shows a Southern transfer of RT-PCR products produced using primers specific for the *bcr/abl* translocation found in CML. The primers and probes used are as described in Kawasaki *et al.* (1988) and are shown in Table 4.2. These primers amplify either a 200 bp product or a 125 bp product depending on whether the translocation involves exon 3 or 2 of the *bcr* gene respectively (Kawasaki *et al.* 1988). The protocol is similar to the one used to identify the *bcl-2* with a few modifications. Messenger RNA is isolated from the samples instead of DNA and the first step of the PCR reaction involves transcription of the mRNA into cDNA. Lanes 2, 4, 5, 6, and 7 are positive with probe C which detects the 200 bp product. Lane 8 is positive with probe D which detects the 125 bp product. Lane 1 contain a digoxigenin-labeled size marker (Boehringer Mannheim). Lane 9 shows the negative result obtained from a water control that is run as a indicator of contamination. The adequacy of the mRNA sample used as the template for lane 3 was confirmed by a positive β-actin PCR (data not shown).

Translocations and Acute Leukemia

In addition to the assay for the *bcr/abl* translocation, RT-PCR assays have been developed for several other clinically significant non-random translocations associated with acute leukemia. The t(1;19), t(4;11) and

Table 4.2. Oligonucleotides used for *bcr/abl* PCR

Name	Sequence
bcr primer	5'-GGAGCTGCAGATGCTGACCAAC-3'
abl primer	5'-TCAGACCCTGAGGCTCAAAGTC-3'
probe C	5'-GCTGAAGGGCTTTTGAACTCTGCTTA-3'
probe D	5'-GCTGAAGGGCTTCTTCCTTATTGATG-3'

Primer and probe sequences are from Kawasaki *et al.* 1988.

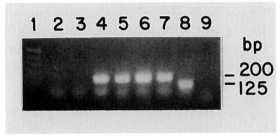

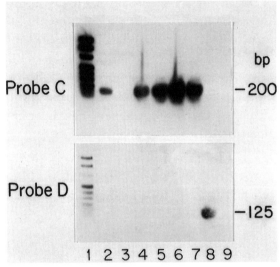

Figure 4.6 Southern blot analysis of RT-PCR products produced using primers specific for the *bcr/abl* translocation found in CML. The upper panel shows an ultraviolet light (UV)-illuminated ethidium bromide stained 2.0% agarose gel. The lower panel shows autoradiographs from the Southern transfer of this gel, hybridized sequentially with probes C and D specific for the 200 bp and the 125 bp PCR products respectively. Lane 1 contains a digoxigenin-labeled size marker. Lanes 2, 4, 5, 6, and 7 are positive with probe C. Lane 8 is positive with probe D. Lane 9 shows the negative result obtained from a water control that is run as a indicator of contamination.

other 11q23 abnormalities are associated with a particularly poor prognosis in pediatric ALL (Ribeiro and Pui 1993). These translocations can be detected using RT-PCR assays (Biondi *et al.* 1993; Borowitz *et al.* 1993; Izraeli *et al.* 1993; Downing *et al.* 1994). Pediatric patients with these translocations should be treated with aggressive regimens designed for patients at high risk for relapse. Translocations detectable by RT-PCR in AML include the t(8;21) and the t(15;17) (Castaigne *et al.* 1992; Chang *et al.* 1992; Chang *et al.* 1993; Downing *et al.* 1993). Both of these translocations are associated with a relatively good prognosis when patients receive specific therapy. The t(15;17) is associated with a specific morphologic subtype of AML, acute promyelocytic leukemia, that responds well to all-*trans*-retinoic acid (RA). RT-PCR assays may be a better predictor of response to RA that conventional cytogenetics or light microscopy

(Miller et al. 1992). Patients with the t(8;21) have a high response rate to chemotherapy when compared with other subtypes of AML. The use of RT-PCR methodologies in the setting of acute leukemia should facilitate the timely stratification of patients into the appropriate treatment protocols and should also be useful for monitoring patients' responses to treatment.

Summary

Molecular biological techniques are playing an increasingly important role in the clinical hematology laboratory, and the importance of these techniques should only increase in the future. Probes for the detection of rearrangements of the immunoglobulin heavy chain and T-cell receptor β chain genes and for the bcr/abl translocation are FDA-approved and are readily available from commercial sources. Advancement in PCR technologies should further enhance the utility of molecular diagnostic testing in the future because it should enable tests to be performed more rapidly, utilizing smaller tissue samples, and with increased sensitivity.

Acknowledgments

The author would like to thank Sandra Prokopetz and Beverly Wood for their technical assistance.

References

Aisenberg, A.C. 1993. Utility of gene rearrangements in lymphoid malignancies. Annu Rev Med 44:75-84.

Alt, F.W., Blackwell, K., Yancopoulos, G.D. 1987. Development of the primary antibody repertoire. Science 238:1079-87.

Ayscue, L.H., Ross, D.W., Ozer, H., Rao, K., Gulley, M.L., Dent, G.A. 1990. Bcr/abl recombinant DNA analysis versus karyotype in the diagnosis and therapeutic monitoring of chronic myeloid leukemia. Am J Clin Pathol 94:404-9.

Benitez, J., Robledo, M. Santon, A., Santos, M., Rivas, C., Echezarreta, G., Castro, P.M. 1992. Correlation between cytogenetic and molecular analysis of t(14;18) in follicular lymphomas. Cancer Genet Cytogenet 59:68-72.

Biondi, A., Rambaldi, A., Rossi, V., et al. 1993. Detection of ALL-1/AF4 fusion transcript by reverse transcription-polymerase chain reaction for diagnosis and monitoring of acute leukemias with the t(4;11) translocation. *Blood* 82:2943-47.

Blennerhassett, G.T., Furth, M.E., Anderson, A., *et al.* 1988. Clinical evaluation of a DNA probe assay for the Philadelphia (Ph1) translocation in chronic myelogenous leukemia. *Leukemia* 2:648-57.

Borowitz, M.J., Hunger, S.P., Carroll, A.J., Shuster, J.J., Pullen, J., Steuber, C.P., Cleary, M.L. 1993. Predictability of the t(1;19)(q23;p13) from surface antigen phenotype: implications for screening cases of childhood acute lymphoblastic leukemia for molecular analysis: a Pediatric Oncology Group study. *Blood* 82:1086-1091.

Castaigne, S., Balitrand, N., de The, H., Dejean, A., Degos, L., Chomienne, C. 1992. A PML/retinoic acid receptor fusion transcript is constantly detected by RNA-based polymerase chain reaction in acute promyelocytic leukemia. *Blood* 79:3110-15.

Chan, L.C., Karhi, K.K., Rayter, S.I., *et al.* 1987. A novel *abl* protein expressed in Philadelphia chromosome positive acute lymphoblastic leukaemia. *Nature* 325:635-37.

Chang, K-S., Fan, Y-H., Stass, S.A., Estey, E.H., Wang, G., Trujillo, J.M., Erickson, P., Drabkin, H. 1993. Expression of AML1-ETO fusion transcripts and detection of minimal residual disease in t(8;21)-positive acute myeloid leukemia. *Oncogene* 8:983-8.

Chang, K-S., Lu, J., Wang, G., *et al.* 1992. The t(15;17) breakpoint in acute promyelocytic leukemia cluster within two different sites of the *myl* gene: targets for the detection of minimal residual disease by the polymerase chain reaction. *Blood* 79:4554-8.

Cheng, G., McLeish, W., Huebsch, L., *et al.* 1992. Rearrangement of BCR genes in malignant lymphoma. *Leukemia* 6:553-55.

Cleary, M.L., Galili, N., Sklar, J. 1986. Detection of a second t(14;18) breakpoint cluster region in human follicular lymphomas. *J Exp Med* 164:315-20.

Cleary, M.L., Smith, S.D., Sklar, J. 1986. Cloning and structural analysis of cDNAs for *bcl-2* and a hybrid *bcl-2*/immunoglobulin transcript resulting from the t(14;18) translocation. *Cell* 47:19-28.

Cleary, M.L., Warnke, R., Sklar, J. 1984. Monoclonality of lymphoproliferative lesions in cardiac-transplant recipients. *New Engl J Med* 310:477-
482.

Cossman, J., Zehnbauer, B., Garrett, C.T., *et al.* 1991. Gene rearrangements in the diagnosis of lymphoma/leukemia. *Am J Clin Pathol* 95:347-
54.

Dhingra, K., Kurzrock, R., Kantarjian, H., *et al.* 1992. Minimal residual disease in interferon-treated chronic myelogenous leukemia: Results and
pitfalls of analysis based on polymerase chain reaction. *Leukemia* 6:754-
60.

Downing, J.R., Head, D.R., Curcio-Brint, A.M., *et al.* 1993. An *AML1/ETO*
fusion transcript is consistently detected by RNA-based polymerase
chain reaction in acute myelogenous leukemia containing the
(8;21)(q22;q22) translocation. *Blood* 81:2860-65.

Downing, J.R., Head, D.R., Raimondi, S.C., *et al.* 1994. The der(11)-
encoded MLL/AF-4 fusion transcript is consistently detected in
t(4;11)(q21;q23)-containing acute lymphoblastic leukemia. *Blood*
83:330-5.

Farkas, D. H. 1993. Quality control of the B/T cell gene rearrangement test.
In *Molecular Biology and Pathology. A Guidebook for Quality Control*
edited by Farkas, D.H., New York: Harcourt Brace Jovanovich, pp. 77-
101.

Frizzera, G. 1992. Atypical lymphoproliferative disorders. In *Neoplastic
Hematopathology*, edited by Knowles, D.M., Baltimore: Williams &
Wilkins, pp. 459-96.

Gale, R.P., Canaani, E. 1984. An 8-kilobase abl RNA transcript in chronic
myelogenous leukemia. *Proc Natl Acad Sci USA* 81:5648-52.

Gribben, J.G., Freedman, A.S., Neuberg, D., *et al.* 1991. Immunologic
purging of marrow assessed by PCR before autologous bone marrow
transplantation for B-cell lymphoma. *N Engl J Med* 325:1525-33.

Gribben, J.G., Neuberg, D., Freedman, A.S., *et al.* 1993. Detection by
polymerase chain reaction of residual cells with the bcl-2 translocation
is associated with increased risk of relapse after autologous bone marrow
transplantation for B-cell lymphoma. *Blood* 81:3449-57.

Gribben, J.G., Saporito, S. Barber, M., *et al.* 1992. Bone marrows of non-Hodgkin's lymphoma patients with a bcl-2 translocation can be purged of polymerase chain reaction-detectable lymphoma cells using monoclonal antibodies and immunomagnetic bead depletion. *Blood* 80:1083-89.

Griesser, H., Tkachuk, D., Reis, M.D., Mak., T.W. 1989. Gene rearrangements and translocations in lymphoproliferative diseases. *Blood* 73:1402-15.

Gulley, M.L., Dent, G.A., Ross, D.W. 1992. Classification and staging of lymphoma by molecular genetics. *Cancer* 69:1600-6.

Gupta, R.K., Whelan, J.S., Lister, T.A., Young, B.D., Bodmer, J.G. 1992. Direct sequence analysis of the t(14;18) chromosomal translocation in Hodgkin's disease. *Blood* 79:2084-88.

Heisterkamp, N., Groffen, J. 1991. Molecular insights into the Philadelphia translocation. *Hematol Pathol* 5:1-10.

Henni, T., Gaulard, P., Divine, M., *et al.* 1988. Comparison of genetic probe with immunophenotype analysis in lymphoproliferative disorders: a study of 87 cases. *Blood* 72:1937-43.

Izraeli, A. Janssen, J.W.G., Haas, O.A., *et al.* 1993. Detection and clinical relevance of genetic abnormalities in pediatric acute lymphoblastic leukemia: a comparison between cytogenetic and polymerase chain reaction analyses. *Leukemia* 7:671-678.

Kawasaki, E.S., Clark, S.S., Coyne, M.Y., Smith, S.D., Champlin, R., Witte, O.N., McCormick, R.P. 1988. Diagnosis of chronic myeloid and acute lymphocytic leukemias by detection of leukemia-specific mRNA sequences amplified *in vitro*. *Proc Natl Acad Sci USA* 85:5698-5702.

Knowles, D.M. 1989. Immunophenotypic and antigen receptor gene rearrangement analysis in T cell neoplasia. *Am J Pathol* 134:761-85.

Knowles, D.M., Athan, E., Ubriaco, A., *et al.* 1989. Extranodal noncutaneous lymphoid hyperplasias represent a continuous spectrum of B-cell neoplasia: Demonstration by molecular genetic analysis. *Blood* 73:1635-45.

Korsmeyer, S.J. 1993. Programmed cell death: bcl-2. In *Important Advances in Oncology 1993*, edited by DeVita, Jr., V.T., Hellman, S., and Rosenberg, S.A. Philadelphia: J.B. Lippincott Company, pp. 19-28.

Kuppers, R., Zhao, M., Rajewsky, K., Hansmann, M.L. 1993. Detection of clonal B cell populations in paraffin-embedded tissues by polymerase chain reaction. *Am J Pathol* 143:230-39.

Kurzrock, R., Gutterman, J.U., Talpaz, M. 1988. The molecular genetics of Philadelphia chromosome-positive leukemias. *N Engl J Med* 319:990-98.

Lee, M.S., Chang, K.S., Freireich, E.J., *et al.* 1988. Detection of minimal residual *bcr/abl* transcripts by a modified polymerase chain reaction. *Blood* 72:893-97.

Limpens, J., de Jong, D., van Krieken, J.H.J.M., *et al.* 1991. *bcl-2/JH* rearrangements in benign lymphoid tissues with follicular hyperplasia. *Oncogene* 6:2271-76.

Loughran, Jr., T.P. 1993. Clonal diseases of large granular lymphocytes. *Blood* 82:1-14.

Medeiros, L.J., Bagg, A., Cossman, J. 1992. Application of molecular genetics to the diagnosis of hematopoietic neoplasms. In *Neoplastic Hematopathology*, edited by Knowles, D.M., Baltimore: Williams & Wilkins, pp. 263-98.

Miller, Jr., W.H., Kakizuka, A., Frankel, S.R., Warrell, Jr., R.P., DeBlasio, A., Levine, D., Evans, R.M., Dimitrovsky, E. 1992. Reverse transcription polymerase chain for the rearranged retinoic acid receptor clarifies diagnosis and detects minimal residual disease in acute promyelocytic leukemia. *Proc Natl Acad Sci USA* 88:2694-98.

Nowell, P.C., Hungerford, D.A. 1960. A minute chromosome in chronic granulocytic leukemia. *Science* 132:1497-99.

Pezella, F., Jones, M. Ralfkiaer, E., Ersbøll, J., Gatter, K.C. Mason, D.Y. 1992. Evaluation of *bcl-2* protein expression and 14;18 translocation as prognostic markers in follicular lymphoma. *Br J Cancer* 65:87-89.

Price, C.G.A., Meerabux, J., Murtagh, S., *et al.* 1991. The significance of circulating cells carrying t(14;18) in long remission from follicular lymphoma. *J Clin Oncol* 9:1527-32.

Pugh, W.C., Pearson, M.C., Vardimon, J.W., Rowley, J.D. 1985. Philadelphia chromosome-negative chronic myelogenous leukaemia: a morphological reassessment. *Br J Haematol* 60:457-67.

Ribeiro, R.C., Pui, C-H. 1993. Prognostic factors in childhood acute lymphoblastic leukemia. *Hematol Pathol* 7:121-42.

Rowley, J.D. 1973. A new consistent chromosome abnormality in chronic myelogenous leukaemia identified by quinacrine fluorescence and Giemsa staining. *Nature* 243:290-93.

Segal, G.H., Scott, M., Jorgensen, T. Braylan, R.C. 1994. Primers frequently used for detecting the t(14;18) major breakpoint also amplify Epstein-Barr viral DNA. *Diagn Mol Pathol* 3:15-21.

Segal, G.H., Wittwer, C.T., Fishleder, A.J., Stoler, M.H., Tubbs, R.R., Kjeldsberg, C.R. 1992. Identification of monoclonal B-cell populations by rapid cycle polymerase chain reaction. *Am J Pathol* 141:1291-97.

Sklar, J. 1992. Antigen receptor genes: Structure, function, and techniques for analysis of their rearrangements. In *Neoplastic Hematopathology*, edited by Knowles, D.M., Baltimore: Williams & Wilkins, pp. 215-244.

Strominger, J.L. 1989. Developmental biology of T cell receptors. *Science* 244:943-50.

Tonegawa, S. 1983. Somatic generation of antibody diversity. *Nature* 302:575-81.

Trainor, K.J., Brisco, M.J., Wan, J.H., Neoh, S., Grist, S., Morley, A.A. 1991. Gene rearrangement in B- and T-lymphoproliferative disease detected by polymerase chain reaction. *Blood* 78:192-96.

Travis, L.B., Pierre, R.V., DeWald, G.W. 1986. Ph1-negative chronic granulocytic leukemia: A nonentity. *Am J Clin Pathol* 85:186-93.

Trumper, L.H., Brady, G., Bagg, A., et al. 1993. Single-cell analysis of Hodgkin and Reed-Sternberg cells: Molecular heterogeneity of gene expression and p53 mutations. *Blood* 81:3097-115.

Tsujimoto, Y., Finger, L.R., Yunis, J., Nowell, P.C., Croce, C.M. 1984. Cloning of the chromosome breakpoint of neoplastic B cells with the t(14;18) chromosome translocation. *Science* 226:1097-99.

Weiss, L.M., Warnke, R.A., Sklar, J., Cleary, M.L. 1987. Molecular analysis of the t(14;18) chromosomal translocation in malignant lymphomas. *N Eng J Med* 317:1185-89.

Williams, M.E., Innes, Jr., D.J., Borowitz, M.J., et al. 1987. Immunoglobulin and T cell receptor gene rearrangements in human lymphoma and leukemia. *Blood* 69:79-86.

CHAPTER 5

Clinical Molecular Genetics

Ruth A. Heim and Lawrence M. Silverman

Introduction

Virtually any disease is the result of an interaction between environmental and genetic factors. The relative contribution of these factors may vary; but it is those disorders known to be caused wholly or partly by inherited or acquired genetic mutations that are the subject of this section. Everyone carries a burden of deleterious genes (perhaps as many as seven, or as few as one), but often two copies of the "disease" gene are needed for a disorder to be expressed. A diagnostic test that identifies individuals who carry one or two copies of a disease-associated gene can provide information about carrier-, risk- or disease-status, thus providing an opportunity to prevent or delay the onset of illness, to treat the early stages of disease more effectively, or to make reproductive decisions based on the test results.

Many genetic tests are based on chromosome analysis or on the activity or concentration of enzymes, serum proteins, or metabolites in blood or urine. For the majority of known genetic disorders, however, diagnosis has relied on clinical features and inheritance patterns, in the absence of other pathological information. In the last decade recombinant DNA techniques have been applied to identifying disease-causing genes with success, leading to a rapid increase in the rate at which genes associated with human disease are reported. Enough progress has now been made that recombinant DNA techniques can be applied to identifying carriers of or individuals at risk for certain genetic disorders, even when the physiological basis for the disorder remains unknown.

In this chapter we describe the current ability to test for a subset of potentially disease-causing genes in the clinical laboratory, and attempt to transmit some of the excitement of recent and prospective advances in

79

understanding molecular genetics. The chapter introduces the genetics of inherited disorders, principles of molecular genetics and their application to diagnostic testing, the kinds of tests currently available, and some of the diseases currently amenable to testing. It should serve as an introduction to more detailed presentations of the molecular pathology of specific genetic diseases in the following chapters.

Medical Genetics

Types of Genetic Disorders

Genetic disorders can be classified in terms of complexity: they may be due to the specific effect of mutations in one gene (single-gene disorders), to the combined effects of mutations in more than one gene (polygenic disorders), or to a change in chromosome number or structure affecting many contiguous genes (chromosomal disorders). When the inherited traits are determined by multiple factors, genetic and possibly environmental, they are said to be inherited as multifactorial disorders.

Single-gene defects may occur in genes located on nuclear chromosomes or on the mitochondrial chromosome, leading to disorders which are usually inherited in a characteristic way. Most single-gene disorders are rare, with a frequency that may be as high as 1 in 500 but is usually much less. The disorders are regularly catalogued in McKusick's *Mendelian Inheritance in Man* under categories defined by the classical patterns of single-gene inheritance—autosomal dominant, autosomal recessive, and X-linked inheritance. Factors influencing inheritance patterns, including germline mosaicism, genomic imprinting, and events leading to uniparental disomy, are increasingly being recognized for their role in disease and in normal development. In particular, interpreting DNA diagnostic tests depends on an understanding of a disorder's inheritance, as will become clear in the following chapters. In 1992, McKusick's catalogue described about 5000 single gene phenotypes, of which at least 3000 are known to be inherited.

For more than 500 single-gene disorders, a basic biochemical defect has been recognized. Disorders such as albinism or phenylketonuria were among the first to be identified as inborn errors of metabolism, by Garrod in 1902. Many of these can be diagnosed by biochemical analysis even without knowledge of which gene is associated with the disorder. The subject of biochemical diagnosis of genetic diseases is not within the scope of this book, but molecular analysis of several of these genetic disorders is now possible. Those disorders for which the disease-causing gene has been identified and cloned (such as the phenylalanine hydroxylase gene defective in phenylketonurics) are also amenable to the diagnostic strategies described in this chapter.

Multifactorial disorders may be congenital malformations, such as neural tube defects, or common adult diseases, such as atherosclerosis, diabetes mellitus, some forms of cancer, and neurodegenerative disorders such as Alzheimer's disease. These disorders tend to recur in families but do not show the characteristic inheritance patterns of single-gene traits. Progress has been made towards identifying altered genes underlying multifactorial disorders, as well as towards identifying associated environmental risk factors, but DNA diagnostic tests for multifactorial disorders are only beginning to be developed in the clinical laboratory. Most clinical applications today are for single-gene defects.

Chromosomal disorders are recognized as an excess or deficiency of all the genes contained within a whole chromosome or chromosome segment. These disorders affect about 7 in 1000 livebirths and account for about half of all spontaneous first-trimester abortions. Traditionally, cytogeneticists analyze chromosome anomalies under the light microscope, using elegant staining techniques. A new field of molecular cytogenetics has developed in the past few years, in which molecular probe technology has been combined with more conventional cytogenetic analyses to develop the technique of fluorescence *in situ* hybridization (FISH). This technique is mainly experimental today, but has the potential to play a significant role in the diagnosis of genetic disease in the molecular pathology laboratory.

Indications for DNA Diagnostic Testing

DNA diagnostic testing is usually provided to two broad groups of individuals: those who are referred to genetic health professionals because of concerns often based on prior knowledge of familial genetic conditions, and those who participate in genetic screening programs. The current paradigm in medical genetics is to provide genetic counseling both to individuals diagnosed with a genetic condition and to their extended families. Genetic information is of consequence not only to the primary consultants and their families, but also to social, health and educational services. Other issues that differentiate the impact of genetic information from most other medical diagnostic information include confidentiality and possible discrimi-

nation, for example, in the workplace or from insurance companies. Counseling, social and ethical issues relating to genetic diagnoses are addressed
in chapter 9.

Common situations that lead to individuals seeking DNA diagnostic testing are listed in Table 5.1. The person seeking genetic information is usually healthy but may wish to know their own status or that of future offspring
or of a current pregnancy. For couples seeking genetic testing one of the goals
may be to prevent recurrence of the genetic disorder in question. Prenatal diagnosis may be available and elective abortion is an option for many
couples. Alternative measures are also available for prevention of recurrence,
including contraception or sterilization, adoption, artificial insemination,
and pre-implantation diagnosis of embryos. For individuals who are diagnosed with a genetic disorder, early treatment may be an advantage (for
example, for phenylketonurics), although many disorders cannot be treated
(for example, Huntington disease). Early warning may be of value to some
individuals. As genotype-phenotype correlations are determined, direct
analysis of mutations associated with a disease may provide information
about the course of the disease.

Individuals tested for genetic disorders are subject to the limitations of
the tests, which in turn are dictated by the available technology and what
is known about the particular gene involved. Several excellent texts,
including Thompson et al. (1991), review these subjects. The overview of
molecular genetics given below addresses some of the issues.

General Principles of Molecular Genetics

Genetic Variation and DNA Polymorphisms

The DNA that constitutes the human genome contains about 6–7 bil-
lion base pairs, organized into 23 pairs of chromosomes. This contrasts
sharply with the approximately 6.4 million base pairs of the prokaryotic bacterium E. coli. One of the reasons for the thousand-fold difference in size
between these genomes has to do with varying degrees of apparent redun-
dancy in the human genome, both within and between genes. If a gene is
crudely defined as the sequence of DNA that encodes a protein, then
more than 95% of the DNA is non-coding. This seems to be true for all
eukaryotic organisms. Some scientists have called the non-coding DNA
"junk," but new technology and analytical techniques are beginning to
provide insights into the nature and function of the "extra" DNA.

Table 5.1. Common Risk Factors Leading to DNA Diagnostic Testing

Indication for Testing	Type of Test:
Healthy indivduals with a known or suspected familial history of a genetic disease	Presymptomatic diagnosis Carrier status Suceptibility testing
Affected individuals with a genetic disease	Clinical diagnosis Prenatal diagnosis Pre-implantation diagnosis
Healthy individuals at increased risk for specific genetic diseases, e.g., because of ethnic background or consanguinity	Clinical diagnosis Carrier status
Newborn infants or adults at increased risk for genetic diseases common in the general population	Clinical diagnosis Carrier status
Healthy individuals or couples who have a reproductive history of one or more children affected with a genetic disease	Carrier status Pre-implantation diagnosis Prenatal diagnosis
Healthy individuals with increased reproductive risks, e.g., because of carrier status or increased maternal age	Prenatal diagnosis Pre-implantation diagnosis

Within every human genome there is a substantial amount of variation, most or all of which is not deleterious. A key concept in genetics is the definition of the kind of variation called polymorphic, because polymorphic variation is used to distinguish individuals, genes and gene products, and to define populations. When two or more variants are present in the population, neither of which is rare, the variants are said to be polymorphic. Each variant must occur at a frequency of greater than 1% to be considered polymorphic. At the molecular level, DNA polymorphisms refer to any difference between the base pair sequences of two chromosomes, at some locus. The different sequences at the locus are referred to as alleles, and the combination of alleles at a locus constitutes a genotype. If a gene is polymorphic, therefore having more than one genotype, then there are potentially more than one physiological manifestations (phenotypes) of the gene.

When two alleles are the same, they are said to be homozygous and when different, heterozygous. Based on the phenotype manifested by its allele, a gene can be described as dominant or recessive, the former when the phenotype is manifested in heterozygotes, the latter when it is manifested in homozygotes. It follows that the typical patterns of single-gene inheritance depend on whether the gene maps to one of the 22 autosomes or to a sex chromosome, and on how the phenotype is manifested. Examples of typical pedigrees showing classical Mendelian inheritance patterns, together with examples of those diseases currently amenable to DNA diagnostic testing, are shown in Figures 5.1 (autosomal dominant inheritance), 5.2 (autosomal recessive inheritance), and 5.3 (X-linked inheritance).

DNA polymorphisms were first identified as variations in the expected lengths of genomic DNA sequences previously digested by restriction enzymes that had cut the DNA at specific recognition sites. A "restriction-fragment-length-polymorphism," or RFLP, occurs whenever a restriction enzyme recognition site varies at the same locus between individuals. Identification of an RFLP requires digestion with a specific restriction enzyme (more than 500 of which have been isolated from bacteria and several 100 of which are commercially available), and gel electrophoresis to separate the fragments by size. If the entire genome is digested, then some means of detecting the fragments of interest among the millions of digested fragments of the human genome is also required. This is usually achieved by Southern blotting (see chapter 2).

Once DNA polymorphisms began to be identified in the late 1970s, it was soon realized that they could be used to "mark" sequences of DNA and even identify individual chromosomes. The first clinically significant RFLP identified in humans was associated with the sickle cell gene (Kan and Dozy 1978). Subsequently, the use of DNA markers such as this one, as well as other markers of increasing complexity, has altered the way disease-causing genes are identified and the way genetic diseases are diagnosed. Furthermore, the concept of using polymorphisms as DNA markers (Botstein *et al.* 1980) led directly to the international Human Genome Initiative (see chapter 12), which has as its first goal to "map" the human and other genomes by positioning closely spaced DNA markers along the length of each chromosome.

The usefulness of a DNA marker for providing information for family studies depends on the frequency of heterozygosity of its alleles. RFLPs are usually di-allelic and have limited utility, but more useful hypervariable

Example of a typical pedigree:

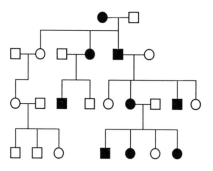

Note: Every child of an affected individual has a 50% risk of inheriting the trait.

Legend: Open circles: unaffected females; Open squares: unaffected males; Filled squares and circles: heterozygous affected males and females.

Selected examples of autosomal dominantly-inherited disorders that can be diagnosed, directly and/or indirectly, in the molecular diagnostic laboratory:

Huntington disease	Marfan syndrome
Neurofibromatosis Types I and II	Multiple endocrine neoplasia
Myotonic dystrophy	Retinitis pigmentosa
Charcot-Marie-Tooth disease 1A and 1B	von Hippel-Lindau syndrome
Familial hypercholesterolemia	Adult Polycystic kidney disease

Figure 5.1. Genetic diseases inherited in an autosomal dominant fashion.

Example of a typical pedigree:

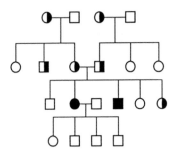

Note: Every sibling of an affected individual has a 25% risk of inheriting the disease.

Legend: Open circles and squares: unaffected non-carrier females and males; Half-filled symbols: unaffected carriers; Filled symbols: homozygous affected individuals

Selected examples of autosomal recessively-inherited disorders that can be diagnosed, directly and/or indirectly, in the molecular diagnostic laboratory:

Cystic fibrosis   Alpha$_1$-antitrypsin deficiency
Tay-Sachs disease   Gaucher disease
Tyrosinase-positive oculocutaneous albinism Sickle cell anemia
Globin disorders   Phenylketonuria

Figure 5.2. Genetic diseases inherited in an autosomal recessive pattern.

Example of a typical pedigree:

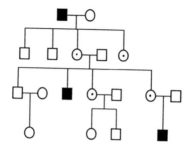

Note: X-linked disorders can be inherited in either a recessive or a dominant fashion. A recessive pattern is shown above. In this example, every male child of a carrier female has a 50% risk of inheriting the disease.

Legend: Open squares and circles: unaffected males and females; Circles containing dots: carrier females; Filled squares: affected males.

Selected examples of X-linked disorders that can be diagnosed, directly and/or indirectly, in the molecular diagnostic laboratory:

Duchenne/Becker muscular dystrophy   Hemophilia A and B
Fragile -X syndrome   Lesch-Nyhan syndrome
Spinobulbar muscular atrophy

Figure 5.3. Genetic diseases inherited in an X-linked pattern.

regions of DNA have been identified in the last decade. These are loci with multiple alleles, each consisting of many copies of the same or similar DNA sequence. VNTR loci (for variable number of tandem repeats), or minisatellites, represent short arrays of a GC-rich sequence which may be from 10–100 bases in length. These are generally present near the ends (telomeres) of chromosomes, and have been used for identity testing by DNA fingerprinting (Jeffreys et al. 1985; Armour and Jeffreys 1992). Simpler repeats, microsatellites, are more generally and more frequently distributed (there are about 50,000 copies per human genome) and constitute di-, tri- and tetra-nucleotide repeat sequences (Weber and May 1989). Microsatellite repeats of the type $(CA)_n$ are probably the most frequently used polymoorphic markers today.

Microsatellites may themselves be the primary cause of genetic disease; to date, seven disorders caused by expansion of a trinucleotide repeat have been identified. These are fragile X mental retardation (see chapter 8), fragile XE mental retardation (Knight et al. 1993), myotonic dystrophy (see chapter 8), spinobulbar muscular atrophy (La Spada et al. 1991), Huntington disease (Huntington's Disease Collaborative Research Group 1993), spinocerebellar ataxia (Duyao et al. 1993), and dentatorubral-pallidoluysian atrophy (DRPLA) (Koide et al. 1994; Nagafuchi et al. 1994). In each case, the number of repeats in unaffected individuals is limited to a specific range. When the number of repeats is expanded beyond the normal range by an unknown mechanism, either the disease or a predisposition to the disease results.

Types of Mutations and Mutation Detection Techniques

For clinical purposes, a mutation can be defined as any permanent change in the DNA that underlies a phenotype; whether it is clinically significant or not depends on the nature of the mutation, its location in the genome, and the tissue(s) involved. Mutations can occur in any cell type, but it is only those in germline cells that can be inherited. Chromosome mutations, in which chromosomes are rearranged or their number is altered, constitute gross mutations that can be detected cytogenetically. Mutations that cannot be seen under the light microscope include base-pair substitutions, DNA sequence rearrangements, such as insertions, deletions and inversions, and certain translocation breakpoints.

Individual genes may have peculiar patterns of mutation, hence posing different diagnostic problems. The authors of the next three chapters provide insights into the practicalities of DNA diagnostic testing for diseases with very different mutation mechanisms and inheritance patterns. For example, point mutations are the most frequent type of change in the human genome. There are more than 350 different point mutations described in the gene which is mutated in cystic fibrosis The molecular pathology of cystic fibrosis is described in chapter 6. By contrast, large deletions are peculiar to the dystrophin gene; about 60% of cases of Duchenne and Becker muscular dystrophy are the result of dystrophin gene deletions. The molecular pathology of Duchenne and Becker muscular dystrophy is described in chapter 7. Lastly, the newly described disease mechanism mentioned above, in which the number of copies of trinucleotide repeat expands above a normal range, is described with respect to the fragile X syndrome and myotonic dystrophy in chapter 8.

Mutation detection techniques are frequently the limiting factor in identifying mutations for diagnostic purposes. Methods can be divided into scanning techniques for identifying previously unknown mutations in DNA sequences, or into techniques for detecting or quantitating known sequence variants. Grompe (1993) provides a good review of mutation detection schemes in use today. Most of the techniques used clinically require prior amplification of the target sequence by the polymerase chain reaction (PCR). Variations on the techniques of PCR analysis, hybridization and transfer technologies, such as Southern blotting, sequence analysis and RNA techniques, were described in chapters 2 and 3. Their applications are elaborated on throughout this book.

DNA Diagnostic Tests in a Clinical Laboratory Setting

Recombinant DNA techniques can be used to detect disease genes in carriers and affected individuals either directly, using specific mutation analysis, or indirectly, using genetic linkage analysis with markers closely linked to or within the gene in question.

Indirect Tests

Once a gene associated with a disease has been mapped, linkage analysis may be the only available, albeit indirect, approach to assessing whether an at-risk individual has inherited the genotype associated with that disease. Similarly, even if a gene has been cloned, if direct mutation analysis is not possible then linkage analysis using intragenic markers may be of value in determining that individual's genotype. Direct analysis is not possible when an individual does not have any of the known mutations tested for, or if mutations have not yet been identified in the gene.

Linkage analysis was the first type of DNA diagnostic test to be used clinically. It is a statistical method, based on following the inheritance of DNA variants or polymorphisms in a family in which a genetic disease is segregating. The method is powerful because it does not require exact knowledge of which specific gene is associated with a disease. Knowledge of map position is sufficient to enable relatively close DNA polymorphisms to be selected for the analysis. Analysis based on closely linked markers is possible because if two loci are close together on the same chromosome, the probability is less that genetic recombination (crossing-over) will occur between the loci when maternal and paternal chromosomes align during meiosis. When no recombination occurs between them, loci are inherited together; this allows one allele of a polymorphic locus to function as a marker for a disease allele. In practice, a "relatively close" marker is no more than 2 centiMorgan (cM) from the gene. A cM refers to genetic distance and is a measure of genetic recombination. At a distance of 2cM, recombination would be detected in 2% of meioses, and this would define the error rate of any linkage analysis performed using a marker at this distance.

Linkage analysis can be Southern-based or PCR-based; any method that identifies variants segregating in a family is acceptable. The key is to identify marker loci sufficiently polymorphic that heterozygous individuals can be seen to transmit different alleles to their offspring. A marker that satisfies this condition in a family is said to be "informative." Linkage analyses and appropriate markers for specific genetic disorders are described in the next three chapters.

Limitations of linkage analysis include the following: errors that arise because of recombination events between the markers used or between the marker and the disease locus; errors because of incorrect assumptions about paternity; difficulties in obtaining appropriate family members; issues of confidentiality among family members; and difficulties in identifying informative markers with sufficient heterozygosity.

Direct Tests

If a mutation is known then it can be detected directly. Direct molecular diagnostic tests require DNA or RNA extracted from an appropriate tissue. In certain cases, tissue-specific genes may be "illegitimately" transcribed at a low level, in peripheral blood lymphocytes, for example, thus providing a source of mRNA that can be reverse-transcribed into cDNA and amplified using the sensitive PCR technique. Using the PCR, it may be possible to design a clinical test that screens cells exfoliated from the respiratory tract, for example, to search for oncogenic mutations.

Standard techniques that directly identify mutations include the PCR, Southern blotting analysis, DNA sequencing, and scanning techniques, such as single-stranded conformation analysis (SSCA) and heteroduplex analysis. These are described when appropriate in each chapter of this book.

Mutations that constitute large deletions and rearrangements, including chromosome translocations, are amenable to analysis by fluorescence *in situ* hybridization (FISH), a technique at the interface of the disciplines of cytogenetics and molecular genetics. This technique has been incorporated into the repertoire of many cytogenetic laboratories, initially as a research tool and more recently for confirmation of cytogenetic or clinical diagnoses. Metaphase chromosome spreads are hybridized with one or more labeled probes and visualized microscopically using one or more fluorochromes and filters to detect fluorescent signals.

FISH probes that have been developed include whole chromosome "paints" that are specific for individual chromosomes. An example of an application of the chromosome 3 paint probe marketed by Oncor is shown in Figure 5.4 (see color figure following page 56). The probe has identified two normal chromosome 3s, as expected, and also a derivative chromosome 22. The derivative 22 contains a normal 22 long arm, but its short arm has been replaced by distal chromosome 3 short arm material. This is diagnostic of an unbalanced translocation, described as 46XX, -22, +der(22)t(3;22) (p24.3;p11).

Other types of FISH probes include unique sequence probes specific for chromosomal regions that, when deleted, are associated with certain phenotypes. An example of the probe to the region of chromosome 15 known to be associated with the Prader-Willi and Angelman syndromes, marketed by Oncor, is shown in Figure 5.5 (see color figures following page 56). Figure 5.5(a) shows a normal chromosome spread in which two loci are identified on each chromosome 15; the distal signal toward the end of each chromosome represents a "marker" that identifies chromosome 15, and the more proximal signal (relative to the centromere) shows that the Prader-Wili/Angelman region is present. In Figure 5.5(b), both chromosome 15 marker signals are present, as is a copy of the Prader-Wili/Angelman probe on one chromosome, but on the other chromosome one copy of the probe is missing. This is diagnostic of a sub-microscopic deletion of chromosome 15: del(15)(q11q13).

FISH has also been successfully applied to identifying extra marker chromosomes, and to detecting specific cancer-associated translocation breakpoints. As the technique continues to be developed, it may well replace slower and more expensive ways of detecting the most common trisomies and sex chromosome aneuploidies. There is also the potential to automate the technique, using combinations of fluorescent tags and fluorochromes.

Automation and Quality Control in the Clinical Laboratory

As new technology emerges, hopes for routine automation of DNA diagnostic tests will be realized. Robots, automated mutation analysis, automated DNA sequencing, simple kits, sequencing chips, and other innovations have become available and are being developed. Several of these developments are described in chapter 2. The additional speed and accuracy provided by automation is an important aspect of quality control for any clinical laboratory. For a molecular diagnostic laboratory, however, another important quality control mechanism, that of regulation and external review, is not as well-developed as it is for clinical laboratories in older disciplines, such as clinical chemistry or cytogenetics.

Medical Genetics was recognized in 1992 as a specialty area by the American Medical Association. Clinical Molecular Genetics is now a certifiable subspecialty of Medical Genetics for which board examinations were first offered in 1993. Regulations and quality control guidelines for clinical molecular genetics laboratories are currently being devised by the American Board of Medical Genetics and also by the College of American Pathologists. At the UNC Hospitals, the molecular genetics laboratory has voluntarily participated in proficiency tests administered by the South-East Regional Genetics Group (SERGG). This proficiency testing fulfills an interim function while uniform guidelines are prepared for laboratories throughout the United States.

Conclusion

Human genome mapping efforts and investigations into the molecular basis of human disease have realized early promises to an extent astonishing even to pioneers in these fields. In 1992, two new major journals were established with a specific focus on human genetic research, emphasizing molecular genetics—*Nature Genetics* and *Human Molecular Genetics*. These publications provide new forums for the expanding scientific output that continues to be published in *The American Journal of Human Genetics* and other well-respected journals.

As potential clinical applications multiply, health-care professionals need to remain alert both to the possibilities and the implications of human genetic research. Many genetic disorders are associated with mental or physical handicaps, or both, and represent a burden on health, social and educational services. Clinical molecular geneticists are already in a position to provide diverse information to patients and professionals; how the information is used continues to evolve.

In the following three chapters (chapters 6–8), the molecular pathology of specific genetic diseases is reviewed. In each case, a particular molecular mechanism underlies a disease with its own peculiar inheritance pattern. To address these differences and still provide accurate diagnostic testing, the techniques used have had to be tailored to each disease. Counseling and social issues arising from DNA diagnostic testing for these and other diseases are addressed in chapter 9.

References

Armour, J.A., Jeffreys, A.J. 1992. Biology and applications of human minisatellite loci. *Curr Opin Genet Dev* 2:850-856.

Botstein, D., White, R.L., Skolnick, M., Davis, R.W. 1980. Construction of a genetic linkage map in man using restriction fragment length polymorphisms. *Am J Hum Genet* 32:314-331.

Duyao, M.P., Ambrose, C.M., Myers, R.H., *et al.* 1993. Trinucleotide repeat length instability and age of onset in Huntington's disease. *Nat Genet* 4:387-392.

Garrod, A.E. 1908. Inborn errors of metabolism. (Lecture II p. 73). *Lancet* ii:1-7.

Grompe, M., 1993. The rapid detection of unknown mutations in nucleic acids. Nature Genet 4:111-117.

Huntington's Disease Collaborative Research Group 1993. A novel gene containing a trinucleotide repeat that is expanded and unstable on Huntington's disease chromosomes. Cell 72:971-983.

Jeffreys, A.J., Wilson, V., Thein, S.L. 1985. Hypervariable 'minisatellite' regions in human DNA. Nature 314:67-73.

Kan, Y.W., Dozy, A.M. 1978. Polymorphism of DNA sequence adjacent to human beta-globin structural gene: Relationship to sickle mutation. Proc Natl Acad Sci (USA) 75:5631-5635.

Knight, S.J., Flannery, A.V., Hirst, M.C., et al. 1993. Trinucleotide repeat amplification and hypermethylation of a CpG island in FRAXE mental retardation. Cell 74:127-34.

Koide, R., Ikeuchi, T., Onodera, O., et al. 1994. Unstable expansion of CAG repeat in hereditary dentatorubral-pallidoluysian atrophy (DRPLA) Nature Genet 6:9-13.

La Spada, A.R., Wilson, E.M., Lubahn, D.B., Harding, A.E., Fischbeck, K.H. 1991. Androgen receptor gene mutations in X-linked spinal and bulbar muscular atrophy. Nature 352:77-79.

McKusick, V.A. 1992. Mendelian inheritance in man. Catalogs of Autosomal Dominant, Autosomal Recessive, and X-linked Phenotypes. 10th edn. Johns Hopkins University Press, Baltimore.

Nagafuchi, S., Yanagisawa. H., Sato, K., et al. 1994. Dentatorubral and pallidoluysian atrophy expansion of an unstable CAG trinucleotide on chromosome 12p. Nature Genet 6:14-18.

Thompson, M.W., McInnes, R.R., Willard, H.F. 1991. Thompson and Thompson. Genetics in Medicine. 5th edn. WB Saunders Co, Philadelphia.

Weatherall, D.J. 1991. The New Genetics and Clinical Practice. 3rd edn. Oxford University Press, Oxford.

Weber, J.L., May, P.E. 1989. Abundant class of human DNA polymorphisms which can be typed using the polymerase chain reaction. Am J Hum Genet 44:388-396.

CHAPTER 6

The Molecular Pathology of Cystic Fibrosis

W. Edward Highsmith, Jr.

Introduction

Cystic fibrosis, or CF, is one of the most common genetic diseases in the Caucasian population. The incidence is approximately 1 in 2000–2500 live births (Boat et al. 1989). CF is present, although much less common, in other ethnic groups; for example, the incidence of CF in the African American population is only 1 in 17,000 births. CF affects many organ systems, with the clinical manifestations resulting from abnormally viscous secretions that obstruct exocrine ducts, reproductive tubules, and airways. The principle morbidity and mortality result from the obstructive lung disease, which is typically characterized by infections with *Staphylococcus aureus* and mucoid *Pseudomonas aeruginosa*. However, CF is a clinically heterogeneous disorder, with some patients having a relatively mild disease and others having a more severe course.

There are several reasons for the current, intense interest in CF. First, the disease is quite common; second, due to advances in therapy, the average age of survival of CF patients has increased from 2–4 years in the 1950s to approximately 30 years in 1993. The effect of this increase in survival means that specialists in all areas of medicine are now likely to encounter patients with CF in their practices. As the first disease gene cloned by a purely positional cloning strategy, and one of the first disease genes to be identified that is associated with a particularly common syndrome, CF is in the position of being a model system for how information accruing from the Human Genome Initiative will be handled. Issues that are being closely watched by the medical community and society at large include issues of

95

broad-based carrier detection programs, patient confidentiality and insurability, and potential costs or savings to the overall health care system. Lastly, with the identification of the defective gene product, research into the pathophysiology of CF has passed a major bottleneck; information on the physiological and biochemical systems involved in the both the normal and disease cases is accumulating at a dizzying pace.

Pathophysiology

The underlying physiological defect giving rise to the observed symptoms is a defect in the transport of ions across epithelial membranes. In CF, there is a decreased secretion of chloride ions and an increased absorption of sodium (Knowles et al. 1983; Boucher et al. 1988; Frizzell et al. 1986; Welsh et al. 1986). These abnormalities of ion transport have been implicated in the pathogenesis of the disease. In the lung, the increased sodium absorption into the epithelial cells lining the airways is accompanied by a net flow of water from the airway secretions into the cells. The normal function of the mucus secretions, which is to trap particulate matter and be cleared by ciliary action, is disrupted as the secretions become dehydrated and increasingly thick and viscous. This sets the stage for colonization by opportunistic bacterial infections. The increased volume of secretions produced due to the inflammatory response is likewise dehydrated. This cycle of events ultimately results in the CF airways being obstructed by an accumulation of a viscous, purulent material which cannot be cleared by the normal cough (Waltner et al. 1987). This scenario is mirrored in the ducts of the pancreas, which normally transport digestive enzymes from the pancreas to the small intestine, with the ultimate result being the complete blockage of the ducts by scarring (Boat et al. 1989). Approximately 85% of CF patients lack sufficient pancreatic function for normal digestion. These patients are referred to as pancreatic insufficient, and are treated with pancreatic enzyme supplementation with every meal (Boat et al. 1989). The inability to resorb sodium and chloride ions in the sweat duct leads to the characteristic elevations in sweat electrolytes. The determination of electrolytes in sweat, particularly chloride, provides an excellent diagnostic test for CF.

Genetics

Cystic fibrosis is inherited as an autosomal recessive trait. In order to express the disease, a zygote must receive two copies of the mutant, or disease gene, one from each parent. Individuals who have one normal gene and one mutant gene do not have the disease or express any symptoms. They can, however, pass the mutant gene to their offspring. These individuals are termed *carriers* of CF. From the incidence of the disease in the Caucasian population it is possible to estimate the frequency of CF carriers in the population using the Hardy-Weinberg equation:

$$p^2 + 2pq + q^2 = 1$$

Where: p^2 is the fraction of the population who are not carriers, i.e., do not have one CF allele.

2pq is the fraction of the population bearing one CF allele, i.e., CF carriers.

q^2 is the fraction of the population bearing two CF alleles, i.e., individuals affected with CF.

From the observed incidence of CF of 1/2500 births:

$$q^2 = 1/2500 \text{ and } q = 1/50$$

Since q is small, $p \sim 1$, and:

$$2pq \sim 2(1)(1/50) = 1/25$$

Thus, approximately one of every 25 Caucasians is a carrier of CF. Figure 6.1 summarizes autosomal recessive inheritance as applied to CF. At a carrier frequency of 1 in 25, the odds of 2 individuals mating, both of whom are carriers, is 1 in 25 times 1 in 25, or 1 in 625. If both parents are carriers of a recessive trait, on average, only 1 child of 4 will express that trait, whereas 2 of 4 children will be carriers, and 1 in 4 will not receive the recessive gene at all. Thus, for CF, and any recessive disorder, if both parents are carriers, the odds are 1 in 4 that any given child will be affected with CF.

Until very recently there has been no test or mechanism to identify CF carriers prior to the birth of an affected child, because a carrier of CF exhibits no symptoms of the disease, and biochemical indices, such as sweat chloride levels, that are useful in the diagnosis of CF, are normal in carriers (Boat *et al.* 1989). Recent advances in the molecular characterization of CF has defined a new role for the clinical laboratory in determination of carrier status in CF families, the general population, and in the prenatal diagnosis of CF in families in which both parents are known to be carriers.

The Cystic Fibrosis Transmembrane Conductance Regulator (CFTR) Gene

Although the physiological defect in CF has been known for several

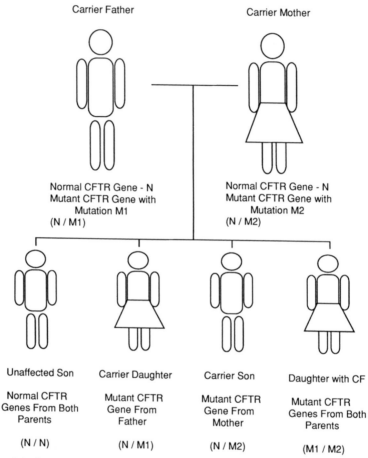

Figure 6.1. Autosomal recessive inheritance of cystic fibrosis.

years, the exact biochemical defect, or the identity of the defective protein, eluded researchers for many years. In 1989, the gene coding for the protein which was defective in CF was identified, not by isolation of the protein by classical biochemical techniques, but by a new approach to the identification of genes termed *positional cloning* (Rommens et al. 1989; Kerem et al. 1989; Riordan et al. 1989).

The CF gene product is termed the cystic fibrosis transmembrane regulatory protein, or CFTR. It is postulated, based on computer modeling and analogy to genes coding for proteins of known structure, to have the structure presented in Figure 6.2. There are twelve regions of the gene which code for stretches of hydrophobic amino acids. Because these regions are just long enough to cross a lipid bilayer, it is predicted that CFTR is a membrane protein with these hydrophobic regions, or transmembrane spanning domains, serving to anchor the protein in the cell membrane. There are two regions which contain Walker motifs (Walker *et al.* 1982) and are predicted to form binding sites for nucleotides, such as ATP, and a large globular domain rich in hydrophilic amino acids and potential phosphorylation sites for protein kinase C. Because chloride transport in respiratory epithelia is known to be regulated by protein kinase C, this region of CFTR is thought to be important in the regulation of its activity, and hence is termed the R domain. The normal function of CFTR

Figure 6.2. Postulated structure of the cystic fibrosis transmembrane conductance regulator.

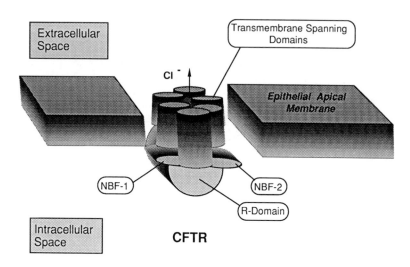

was not obvious from its primary structure; however, substantial evidence has accrued suggesting that it serves as a channel for the transport of chloride ions across the epithelial cell apical membranes (Drumm et al. 1990; Anderson et al. 1991; Berger et al. 1993). However, recent evidence indicates that CFTR may have more than one physiological role. Gabriel et al. have shown that CFTR has a distinct regulatory relationship with the outwardly rectifying chloride channel (Gabriel et al. 1993). It is not yet clear how deficiencies of the two described functions of CFTR account for the observed pathology. Intensive efforts are underway in many laboratories across the world to try and understand the normal function and regulation of CFTR activity and its role in lung defense. Ultimately, this knowledge will assist in the development of therapeutic strategies for modifying or bypassing abnormal CFTR function.

The initial sequence analysis of CFTR genes cloned from a normal individual and an individual affected with CF revealed that the sole difference between the two was the deletion of three base pairs of DNA in the mutant chromosome. This three base pair deletion causes the deletion of a single phenylalanine residue at position 508 of the CFTR protein and is termed ΔF_{508}. Position 508 occurs in the first nucleotide binding fold and presumably is important in ATP binding. The deletion of this amino acid was originally thought to interfere with ATP binding (Rommens et al. 1989; Kerem et al. 1989; Riordan et al. 1989). However, the ΔF_{508} mutation has been shown to cause a cellular trafficking lesion, with the protein being retained in the Golgi apparatus instead of being inserted into the apical membrane (Cheng et al. 1990; Kartner et al. 1992). Interestingly, the effect of the ΔF_{508} mutation has been shown to be temperature sensitive in vitro, with overexpressing cells cultured at 37°C showing the subcellular mislocalization defect, while cells cultured at 27°C show normal placement of the mutant protein into the cell membrane and a corrected ion flux (Denning et al. 1992).

Population Screening

Prior to the actual cloning of the gene, it was anticipated and hoped that CF would prove to be a genetically homogeneous disorder, much like sickle cell anemia, with a single mutation in the disease gene causing all cases of CF. The reason for this hope was that a genetically homogeneous disease is a candidate for a population-based carrier detection program. It was hoped that identification of CF carriers prior to the birth of an affected child could lead to a marked decrease in the incidence of CF much as the incidence of Tay-Sachs disease in the Jewish population and the incidence of thalassemia in the Mediterranean basin has decreased after the introduction of carrier screening and educational programs (Scriver et al. 1984; Sandkoff et al. 1989). Furthermore, as the United States Office of Technology Assessment (OTA) has calculated that there are approximately 100–125 million individuals of child-bearing age in the U.S. who could potentially be screened (OTA 1992); major genetic testing laboratories recognized a huge potential market for their services. The reason for the optimism that CF would prove to be genetically homogeneous was the discovery in late 1987 of a cluster of polymorphisms very close to the CF gene (Estivill et al. 1987a). Two of these, termed KM-19 and XV.2c, were shown to have crossover rates with the CF gene of only 0.1%. When analysis for both of the diallelic XV.2c and KM-19 markers is performed, there are four possible results per chromosome: there can be a "1" allele for XV.2c and a "1" allele for KM-19, etc. Table 6.1 lists these possibilities.

Table 6.1 Distribution of Haplotypes on Normal and CF Chromosomes

Haplotype	XV.2c Allele	KM-19 Allele	% Normal Chromosomes	% CF Chromosomes
A	1	1	30	7
B	1	2	15	86
C	2	1	42	2
D	2	2	13	5

Adapted from Lemna et al. 1990.

The pattern of alleles was defined as a *haplotype* and designated A-D. Some distribution of these haplotypes was observed on normal, non-CF chromosomes. However, the distribution on CF chromosomes was observed to be markedly different, with the "B" haplotype, which is relatively infrequent on normal chromosomes, accounting for 86% of CF chromosomes (Estivill et al. 1987b; Beaudet et al. 1989). The fact that the CF gene and the polymorphic markers were not in genetic equilibrium, in which the distribution would be the same for CF and normal chromosomes, but rather in *genetic or linkage disequilibrium*, was the reason for the optimism that

CF would turn out to be caused by a single gene mutation. Prior to the cloning of the gene itself, it appeared as if at some time in the past, the (only) CF mutation had occurred on a chromosome that by chance carried the "B" haplotype. In fact, as generations passed and the rare crossover between the CF disease gene and the RFLP markers occurred, the disease started appearing on other haplotypes.

Unfortunately, the ΔF_{508} mutation is not the sole genetic lesion causing CF. In the original description of the gene, a frequency of 68% was found for the ΔF_{508} mutation in a population of 107 Canadian CF patients (Kerem et al. 1989). This frequency was confirmed in a subsequent, worldwide collaborative study of over 13,000 CF chromosomes (The CF Gene Analysis consortium 1990). However, the frequency of the most common mutation varied substantially geographically, with the highest frequencies (70–80%) seen in Northern Europe and lower frequencies (30–50%) seen in Southern Europe, Northern Africa, and Israel. In the United States, the frequency of the ΔF_{508} mutation on the chromosomes of CF patients is approximately 75% (Lemna et al. 1990; Highsmith et al. 1990). The inability of a test for a single, disease-causing mutation to identify all carrier individuals has profound effects on the statistics of population-based carrier detection programs. At a 75% detection rate for carriers (75% sensitivity), the post-test probability of a ΔF_{508}-negative individual being a CF carrier is 1 in 100. When this test is applied to couples, there are three possible outcomes: both individuals test negative for ΔF_{508} (Neg. x Neg. couples), both individuals test positive (Pos. x Pos. couples), or one individual tests positive and the other tests negative (Pos. x Neg. couples). Neg. x Neg. couples have a probability of having a child with CF of 1/40,000 (1/100 x 1/100 x 1/4), a probability greatly reduced from the a priori odds of 1/2500 (the incidence)(Galen and Gambino 1975). Approximately 94% of Caucasian couples tested would fall into this category. 56% (75% x 75%) of carrier couples would be identified and could be referred to genetic counseling. However, a Pos. x Neg. couple will have an increased risk of having offspring with CF (1 x 1/100 x 1/4 = 1/400). Approximately 6% of couples tested will fall into this category. The traditional response to a couple at increased risks for genetic disease is genetic counseling and, if possible, prenatal diagnosis. There exists no effective means of offering prenatal diagnosis to a Pos. x Neg. couple. Although the measurement of microvillar intestinal enzymes in amniotic fluid, such as alkaline phosphatase and gamma-glutamyl transferase, have been used for prenatal diagnosis of CF, these tests yield approximately 5% false positive results and 8% false negative results (Beaudet and Buffone 1987). Furthermore, the number of couples who would be faced with these intermediate risks is large, and the number of trained genetic counselors needed to deal with the volume of couples who would need intensive genetic education and counseling is not sufficient. In addition, the genetics community was unsure of the levels of anxiety that Pos. x Neg. couples would face. These factors have led the genetics community to decide that large scale screening programs are inappropriate until either:

1. the number of CF chromosomes that can be identified by direct testing reaches the 90–95% level, or
2. alternative strategies for the delivery of genetic information are developed that can be shown to be as effective in terms of patient education as classical genetic counseling (Beaudet et al. 1990; Caskey et al. 1990).

CFTR Mutations and Population Screening Strategies

Due to the large interest in finding the mutations that account for the remaining 25–30% of CF chromosomes, many laboratories across the world have been involved in the search for new mutations. It was hoped that only a handful of mutations other than ΔF_{508} would prove to be involved with CF. Unfortunately, this hope turned out to be as overly optimistic as the previous notion that CF would be caused by a single mutation. Figure 6.3 is a schematic of the CFTR gene showing the positions and type of over 350 mutations identified and reported to the Cystic Fibrosis Gene Analysis Consortium as of September 1993. There are a number of types of mutations that have been found, including the following: *missense* mutations, or DNA point mutations which cause the code for the proper amino acid to now code for another amino acid; *nonsense* mutations, where the substitution of one nucleotide for another at the DNA level causes the code for an amino acid to read Stop; *frameshift* mutations, in which small insertions or deletions disrupt the reading frame; and *splice site* mutations, in which point mutations in the control regions flanking individual exons disrupt normal splicing of the message. Although there have been mutations found in all areas of the gene, some clustering of mutations of all types is evident in the exons which comprise the nucleotide binding domains. The significance of this clustering is not clear, but it may indicate that

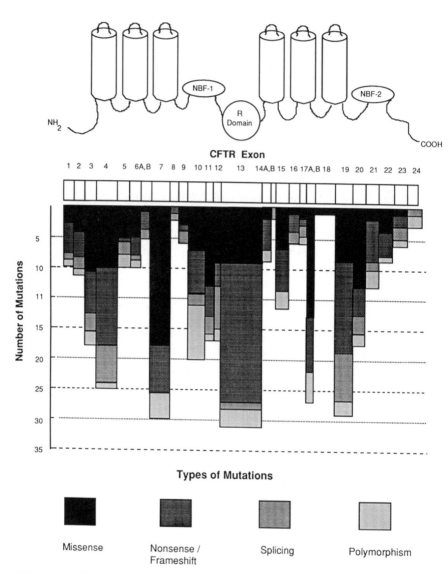

Figure 6.3. Distribution of mutations in the CFTR gene.

the ATP-binding regions are more sensitive to mutations in that any disruption of structure will lead to a non-functional protein. Other areas may be more robust in that they can tolerate some mutations without causing disease. Since mutations are only being sought in CF patients, there is likely to be a bias toward finding mutations that render the protein non-functional.

Of the many mutations described to date, the majority are apparently quite rare, and can almost be considered "private" or "family" mutations. A few, however, are present on 1–2% of CF chromosomes each. Testing for these mutations, N1303K (Osborne et al. 1991), R553X, G551D (Cutting et al. 1990), and G542X (Kerem et al. 1990), increases the frequency of mutations detectable in the United States to approximately 80%. Many large laboratories in the United States are currently testing for 20–30 mutations with detection rates of 85–90%. Other mutations have been described that occur at high frequencies in certain geographic locations or in certain ethnic groups. For example, the frequency of the ΔF_{508} mutation in CF patients of Ashkenazi Jewish extraction is only 30%. Another mutation, a nonsense mutation in the second nucleotide binding fold, W1282X, is found on over 50% of CF chromosomes in this ethnic group (Vidaud et al. 1990; Shoshani et al. 1992). Testing for five mutations, ΔF_{508}, W1282X, G542X, N1303K, and 3489+10 Kb C to T (Highsmith et al. in press), identifies 97% of CF alleles in this population (Abeliovich et al. 1992).

It has become clear that a strategy of testing for more and more mutations will neither reach the 95% detection level nor be economically viable. In response to these observations, three component institutes of the National Institute of Health (the National Center for Human Genome Research, the National Center for Nursing Research, and the National Institute of Child Health and Human Development) announced in September 1991 funding of seven pilot programs for general population CF carrier screening. These studies will fully investigate issues of education, counseling, and the provision of laboratory data. They will also test novel methods of providing genetic information to the general public. Results of the studies are anticipated in 1994–1995.

The NIH studies are testing educational tools to facilitate the public's understanding and acceptance of testing strategies that are less than 100% sensitive. An alternative paradigm was recently proposed, that of a two-stage approach in which an extended panel would only be utilized in Pos. x Neg. couples. Under this scenario, approximately 93% of couples tested would have their needs met by a relatively simple, low-cost procedure designed to detect 80–85% of mutations. Only an estimated 7% of couples would be referred to a more extensive and expensive second-tier screen (Beaudet and O'Brien 1992). Screening techniques, such as single-strand conformational analysis (SSCA)(Orita *et al.* 1989) or denaturing gradient gel electrophoresis (DDGE)(Myers *et al.* 1987) are candidates for this type of second-stage test. One particularly promising technique is heteroduplex analysis. This procedure not only has a very high sensitivity (percent mutations detected), but it also yields patterns characteristic of each mutation (Figure 6.4) (see color insert following page 56). This technique, combined with a multiplex amplification step, is under development in the author's laboratory for a second-tier test for CF carrier detection programs.

Genotype-Phenotype Correlations

The observations of the heterogeneity in the clinical expression of disease and the myriad of disease causing alleles naturally leads to the question of genotype-phenotype correlations. It will be important to correlate the clinical phenotypes with the CFTR genotypes to improve diagnostic/prognostic capabilities and, perhaps more importantly, to provide CFTR structure/function insights. Ultimately, structure/function relationships will contribute to understanding CFTR's role in lung defense and assist in the development of therapeutic strategies for modifying or bypassing abnormal CFTR function.

That the genotype-phenotype relationships would not be straightforward, at least with respect to the major cause of morbidity and mortality, pulmonary disease, was demonstrated by the Toronto group soon after the cloning of the gene. In an analysis of 293 CF patients who had been well characterized with respect to clinical expression of disease and pancreatic function, and who had been genotyped at the ΔF_{508} locus, it was found that there was no statistically significant difference in a standard index of pulmonary function, the Fraction of Expired Volume in One Second (FEV1), between pancreatic insufficient (PI) patients who bore the genotypes $\Delta F_{508}/\Delta F_{508}$, ΔF_{508}/Other, or Other/Other. Of particular note, however, was the extreme variability observed when a plot of FEV1 versus patient age was constructed. At almost any given age, there were patients with FEV1 values approaching normal and patients with dire lung disease. Likewise, there was no apparent difference in the pulmonary status between pancreatic sufficient (PS) patients grouped by genotypes ΔF_{508}/Other or Other/Other. This group was also markedly heterogeneous. However, the PI patients tended to have more severe disease than the PS patients with higher sweat chloride values, earlier age of diagnosis, more severe pulmonary disease for age, and lower percentiles for weight. The authors concluded that there did exist a correlation between genotype and phenotype of pancreatic disease, but that other influences (genetic or environmental) precluded the observation of a correlation for pulmonary disease (Kerem *et al.* 1990). These conclusions confirmed a hypothesis put forward in the original paper describing the cloning of the gene, namely that different CFTR mutations are associated with recognizable differences in disease expression, particularly with respect to pancreatic function. The ΔF_{508} mutation, when present in homozygous dosage, is almost invariably associated with PI and a more severe disease expression. Furthermore, when paired with another "severe" mutation, the result is a severe disease, indistinguishable from that due to ΔF_{508} homozygosity. However, when paired with a "mild" mutation, the result is PS and a milder form of CF (Kerem *et al.* 1989).

Rigorous categorization of other mutations with respect to whether they are "severe" or "mild" has been difficult due to, as predicted by Kerem *et al.* 1989, the relative scarcity of all other mutations than ΔF_{508}. However, as more laboratories began testing for batteries of mutations, certain correlations began to become apparent. Among these was the striking and unexpected finding that, when CF was defined not as a clinical syndrome, but by whether or not an individual had two CFTR mutations, the previously accepted range of what constituted CF was greatly expanded.

In 1992, a group of investigators at Boston University performed CFTR genotyping on a group of 25 males who were seen at a fertility clinic and diagnosed with congenital bilateral absence of the vas deferens (CBAVD). As the carrier frequency of CF is approximately 1 in 25, it would be anticipated that, if there were no genetic relationship between CF and CBAVD, perhaps one individual bearing an identifiable CFTR mutation would be detected in this study. In fact, 16 (64%) of the 25 study individuals had CFTR mutations, 13 of them ΔF_{508}. Two individuals had alanine to glycine mutations (G576A) on the opposite chromosome and one individual had the D1270N mutation. This study expanded our knowledge of the breadth of phenotypic abnormalities that can be associated with CFTR abnormalities and significantly broadened the definition of what constitutes a "mild" mutation (Anguiano *et al.* 1992).

Although the sweat chloride test is an extremely powerful tool for the diagnosis of CF, with approximately 99.9% of CF patients giving a positive test result, it was not uncommon for large CF centers to have one or two patients with clinical features consistent with a mild, typically PS, presentation of CF, but who, on repeated measurements, had sweat chloride values within the reference range (Davis *et al.* 1980; Stean *et al.* 1978; Sarsfield and Davies 1975). Although these patients were typically treated as having CF, they remained a diagnostic enigma until the development of a technique for the measurement of parameters related to ion transport directly in the respiratory epithelia *in vivo*, the nasal potential difference (Knowles *et al.* 1981). Over the course of several years prior to the cloning of the gene, Dr. Michael Knowles at the University of North Carolina at Chapel Hill began collecting and characterizing these patients against the day that genotypic inquiries could be made. Using a RNA-PCR and direct sequencing approach, several novel mutations were identified by the North Carolina group. These mutations correlated with the triad of relatively mild pulmonary disease, pancreatic sufficiency, and normal sweat chloride values. One of the mutations, 3489+10 Kb C>T (Highsmith *et al.* in press), was novel for several reasons. First, it was the first disease-causing point mutation to be found deep inside a large intron. This lends support to the notion that introns serve as reservoirs of genetic information in eukaryotes and serve to increase the rate at which proteins can evolve and assume new functions (Gilbert 1978). Second, it was the first CFTR mutation to be associated with modulation of the amount of normal protein produced. In this regard it is similar to the ß⁺ alleles in the beta-thalessemia in which the disease is caused by an underproduction of beta-globin. These alleles tend to give rise to milder disease than the ß⁻ alleles in which no functional beta-globin is produced. Lastly, it is a relatively common allele, being present not only in approximately half of the CF patients with normal sweat electrolytes in the North Carolina study, but in approximately 5% of CF alleles in the Ashkenazi Jewish population (Abeliovich *et al.* 1992)

With the identification of alleles which give variant presentations of CF, the identification of several mutations associated with PS CF, and a host of mutations associated with PI, a picture of the correlation of genotype and phenotype began to emerge. This is presented in tabular form in Table 6.2.

However, this model, which associates each mutation with a specific phe-

Table 6.2. Genotype-phenotype correlations.

Decreasing Severity of Clinical Disease →

Lung Disease:	Severe	Less Severe	Mild	Absent
Pancreatic Status:	PI	PS	PS	PS
Sweat Chloride:	Elevated	Elevated	Normal	Normal
Vans Defrens:	Absent	Absent	Absent	Absent
Mutation:	ΔF508	R117H	3849+10 Kb C>T	G576A
	G542X	2789+5 G>T	G551S	D1270N
	G551D	R347P	Others	Others
	R553X	R334W		
	N1303K	Others		
	W1282X			
	Many Others			

notype, is not perfect. In the study by Kerem and co-workers (Kerem *et al.* 1990), a patient homozygous for ΔF$_{508}$ was reported to be PS. Although the intron 19 mutation was originally described in patients with normal sweat chloride values, several groups have described patients with elevated sweat chlorides, and even with PI (Highsmith *et al.* in press; Gilbert *et al.* in press; Augarten *et al* 1993).

The boundaries separating the genotypes and phenotypes in Table 6.2 must be considered to represent the general findings only; the results cannot be extrapolated to provide a firm prognosis in an individual patient.

The most rigorous study to appear to date on the variability of clinical findings within a genotype was an international consortium effort, headed by Drs. Hamosh, Cutting (Johns Hopkins School of Medicine) and Corey (Hospital for Sick Children, Toronto). Recognizing the difficulty that any one laboratory faced in generating a statistically significant number of data points for a study on genotype-phenotype correlations in patients with less common genotypes, 32 laboratories worldwide agreed to pool their data on patients with the more common non-ΔF_{508} mutations. Utilizing an elegant study design and statistical analysis, the study concluded that patients who are compound heterozygotes with ΔF_{508} and one of the most common non-ΔF_{508} mutations (G542X, R553X, W1282X, N1303K, 621+1 G>T, or 1717-1) are indistinguishable from ΔF_{508} homozygotes using any of the clinical parameters investigated. The only exception was, as expected, patients who had the genotype R117H/ΔF_{508} or R117H/R117H, who were clearly different from the age- and sex-matched $\Delta F_{508}/\Delta F_{508}$ controls. The majority of the patients bearing the R117H mutation were PS while the control group was uniformly PI. The R117H patients tended to be diagnosed at a later age and had lower sweat chloride values. Interestingly, however, the FEV1 values and the Shwachman overall clinical scores were not statistically different from the controls (The CF Genotype-Phenotype Consortium 1993).

The study confirmed the hypothesis proposed by Kerem et al. (1989) that genotype would be a predictor of pancreatic function. It also demonstrated that, due to the large variability within each group, pulmonary function was not predicted by genotype, and that other influences, either genetic, environmental or both, affect the pulmonary phenotype (The CF Genotype-Phenotype Consortium 1993). Thus, this study is in agreement with two other international collaborative studies with examined the genotype-phenotype relationships of the mutations N1303K (Osborne et al. 1992) and G551D (Hamosh et al. 1992).

Family Studies

Direct Analysis

Although implementation of wide-scale carrier detection programs is being held pending the results of the NIH studies, the search for mutations and the efforts to ascertain their frequency has provided valuable dividends for families in which CF has already occurred. Relatives of CF patients are often concerned regarding their own carrier status, and the parents of a CF child often want to ensure that any further children are not affected. The ability to detect approximately 80–85% of CF chromosomes provides these families with a valuable service. In preparation for doing family studies, a large effort to obtain a blood sample from the CF patient should be made, if possible, as identification of the mutations in the proband allows the laboratory to perform a directed search in other family members. However, it is important not to neglect the potential for other mutations being present in the family. Only when a given chromosome can be traced by pedigree analysis to a demonstrated carrier (i.e., an individual who has a known mutation on the opposite chromosome, yet is free of disease), can that allele be considered proven to be risk-free. Many laboratories screen the patient sample for the most common CF mutation, ΔF_{508}, as in approximately 50% of cases, the affected individual, or proband, will prove to be homozygous for the ΔF_{508} mutation. In cases in which the proband is not homozygous for ΔF_{508}, other mutations can be tested for. Alternatively, many other laboratories are performing semi-automated procedures for the detection of multiple mutations, often 20 or more (Shuber et al. 1993). In families in which both mutations are identifiable, it is quite straightforward to do direct mutation detection on any of the patient's relatives and ascertain whether or not they are carriers of CF. Similarly, it is straightforward to perform the assay on fetal tissue obtained by amniocentesis or chorionic villus sampling, thereby diagnosing or ruling out CF prenatally.

Indirect Analysis

In cases in which only one or no causative mutations are identified in the affected individual, a direct test to determine carrier status in all of the relatives is not applicable. In these cases indirect testing or linkage analysis must be employed. Indirect analysis must be used when the precise nature of the molecular defect is not known. In this type of analysis, the mutation-bearing chromosome is identified, not by identification of the mutation, but by co-inheritance with restriction fragment length polymorphisms (RFLP's) or variable number of tandem repeat polymorphisms (VNTR's)(Highsmith and Friedman 1993). Indirect analysis has several disadvantages. The primary one is that it depends on the existence of at least one affected child in the family in order to determine which of the four parental chromosomes harbors the disease gene. Furthermore, there must be an unequivocal diagnosis in the affected individual. Obviously, if the diagnosis of a genetic disease is incorrect, then the laboratory will not be able to trace the actual disease gene in the family, with the possibility that individuals will make reproductive decisions based on incorrect information. In addition, multiple members of the family must be studied. Of crucial importance is correct paternity; it is apparent that if the analysis is performed with an incorrect assumption regarding the biological father, it is impossible to interpret the results correctly. There is also an error rate associated with indirect analysis due to meiotic recombination or crossing-over. The magnitude of this error rate is determined by the distance between the polymorphic site and the disease gene. The final difficulty with indirect or linkage testing is the problem of informativity, as the frequency of informative meioses increases with increasing heterozygosity of the polymorphism. Although linkage analysis is complicated by many more factors than direct analysis, given a careful family history, cooperation of all necessary family members, and probes close to the disease gene, indirect analysis is an extremely useful tool in the molecular genetics laboratory.

There are several RFLP's on chromosome 7 near or within the CFTR gene (White et al. 1985; Wainwright et al. 1985; Beaudet et al. 1986; Dörk et al. 1992; Zielenski et al. 1991; Highsmith 1993). Fortunately, the combination of the availability of intragenic polymorphisms, with negligible risk of error due to recombination, and highly informative CA repeats, reduces the difficulties with linkage analysis. Using the currently available polymorphisms, virtually all CF families are informative.

Conclusion

The rate at which our fundamental understanding of the ion physiology, molecular genetics, and pathophysiology of one of the most common, life-shortening inherited diseases of human-kind, cystic fibrosis, has grown over the past five years has given hope to many families in which this disease has occurred. Furthermore, the dizzying pace of CF research bodes well for eventual understanding of the genetic basis for many disease states, both mono- and poly-genic. The clinical laboratory has seen its role undergo massive and rapid change. Today, the laboratory is able to provide information to large groups of consumers, e.g., carriers of genetic diseases, that it was not able to assist previously. Furthermore, as the correlations between genotype and phenotype become more clear, both for CF and other diseases, the laboratory will be able to provide vastly improved prognostic projections. The clinical laboratory stands at the junction between the research laboratory and clinical practice. The coming decade will be that of the gene; it will be the challenge for the clinical laboratory to utilize its long history of making the transition from research to practice in order to continue to deliver quality health care services to those in need.

References

Abeliovich, D., Lavon, I.P., Leher, I., Cohen, T., Springer, C., Avital, A., Cutting, G. 1992. Screening for five mutations detects 97% of cystic fibrosis (CF) chromosomes and predicts a carrier frequency of 1:29 in the Jewish Ashkenazi population *Am J Hum Gen* 51:951-6.

Anderson, M.P., Gregory, R.J., Thompson, S., *et al.* 1991. Demonstration that CFTR is a chloride channel by alteration of its anion selectivity. *Science* 253:202-5.

Anguiano, A., Oates, R.D., Amos, J., *et al.* 1992. Congenital bilateral absence of the vas deferens - A primarily genital form of cystic fibrosis. *JAMA* 267:1794-7.

Augarten, A., Kerem, B-S., Yahav, Y., *et al.* 1993. Mild cystic fibrosis and normal or borderline sweat test in patients with the 3849+10 kb C to T mutation. *Lancet* 342:25-6.

Beaudet, A.L., Bowcock, A., Buchwald, M., *et al.* 1986. Linkage of cystic fibrosis to two tightly linked DNA markers: joint report from a collaborative study. *Am J Hum Gen* 39:681-93.

Beaudet, A.L., Buffone, G.J. 1987. Prenatal diagnosis of cystic fibrosis. *J Ped* 111:630-3.

Beaudet, A.L., Feldman, G.L., Fernbach, S.D., Buffone, G.L., O'Brien, W.E. 1989. Linkage disequilibrium, cystic fibrosis, and genetic counseling. *Am J Hum Gen* 44:319-26.

Beaudet, A.L. 1990. Carrier screening for cystic fibrosis [editorial] *Am J Hum Gen* 47:603-5.

Beaudet, A.L., O'Brien, W.E. 1992. Advantages of a two-step laboratory approach for cystic fibrosis carrier sceening. *Am J Hum Gen* 50:439-40.

Boucher, R.C., Cotton, C.U., Gatzy, J.T., Knowles, M.R., Yankaskas, J.R. 1988. Evidence for reduced Cl- and increased Na+ permeability in cystic fibrosis human primary cell cultures. *J Physiol (Lond)* 405:77-103.

Boat, T.F., Welsh, M.J., Beaudet, A.L. 1989. Cystic fibrosis. In Scriver CL,

Beaudet AL, Sly WS, Valle D, eds. *The metabolic basis of inherited disease*, 6th ed. New York: McGraw-Hill, 2649-80.

Caskey, C.T., Kaback, M.M., Beaudet AL, *et al.* 1991. The American Society of Human Genetics statement on cystic fibrosis screening. *Am J Hum Gen* 47:608-12.

Cheng, S.H., Gregory, R.J., Marshall, J. *et al.* 1990. Defective intracellular transport and processing of CFTR is the molecular basis of most cystic fibrosis. *Cell* 63:827-34.

Congress of the the United States of America, Office of Technology Assessment. 1992. *Cystic fibrosis and DNA tests: Implications of carrier screening.* U.S. Government Printing Office, Washington, D.C. pp 49-66, 141-166.

Cutting, G.R., Kasch, L.M., Rosenstein, B.J., *et al.* 1990. A cluster of cystic fibrosis mutations in the first nucleotide-binding fold of the cystic fibrosis conductance regulator protein. *Nature (London)* 346:366-9.

The Cystic Fibrosis Genetic Analysis Consortium. Worldwide survey of the ΔF508 mutation—Report from the CF Genetic Analysis Consortium. 1990. *Am J Hum Gen* 47:354-9.

The Cystic Fibrosis Genotype-Phenotype Consortium. 1993. Correlation between genotype and phenotype in cystic fibrosis patients. *New Eng J Med* 329:1308-13.

Davis, P.M., Hubbard, V.S., Di Sant'Agnese, P.A. 1980. Low sweat electrolytes in a patient with cystic fibrosis. *Am J Med* 69:643-6.

Denning, G.M., Anderson, M.P., Amara, J.F., Marshall, J., Smith, A.E., Welsh, M.J. 1992. Processing of mutant cystic fibrosis transmembane conductance regulator is temperature sensitive. *Nature (Lond)* 358:761-4.

Dörk, T., Neumann, T., Wulbrand, U., *et al.* 1992. Intra- and extragenic marker haplotypes of CFTR mutations in cystic fibrosis families. *Hum Genet* 88:417-25.

Drumm, M.L., Pope, H.A., Cliff, W.H., *et al.* 1990. Correction of the cystic fibrosis defect *in vitro* by retrovirus-mediated gene transfer. *Cell* 62:1227-33.

Estivill, X., Farrall, Scambler, P.J., *et al.* 1987. A candidate for the cystic fibrosis locus isolated by selection for methylation-free islands. *Nature (London)* 326:840-5.

Estivill, X., Scambler, P.J., Wainwright, B.J., *et al.* 1987. Patterns of polymorphism and linkage disequilibrium for cystic fibrosis. *Genomics* 1:257-63.

Frizzell, R.A., Rechkemmer, G., Shoemaker, R.L. 1986. Altered regulation of airway epithelial cell chloride channels in cystic fibrosis. *Science* 233;558-60.

Gabriel, S.E., Clarke, L.L., Boucher, R.C., Sutts, M.J. 1993. CFTR and outward rectifying chloride channels are distinct proteins with a regulatory relationship. *Nature* 363:263-6.

Galen, R.S., Gambino, S.R. 1975. *Beyond normality: The predictive value and efficiency of medical diagnosis.* New York: John Wiley and Sons.

Gilbert, W. 1978. Why genes in pieces? *Nature (Lond)* 271:501.

Gilbert, F., Li, Z., Bialer, M., *et al.* Genotype-phenotype correlations in cystic fibrosis: Lessons from mutation 3849+10 kb C to T. *Am J Hum Gen* (in press).

Hamosh, A., Kig, T.M., Rosenstein, B.J., *et al.* 1992. Cystic fibrosis patients bearing both the common missense mutation gly to asp at codon 551 and the ΔF508 mutation are clinically indistinguishable from ΔF508 homozygotes, except for a decreased risk of meconium ileus. *Am J Hum Gen* 51:245-50.

Highsmith, W.E., Chong, G.L., Orr, H.T., *et al.* 1990. Frequency of the delta Phe508 mutation and correlation with XV.2c/KM-19 haplotypes in an American population of cystic fibrosis patients: Results of a collaborative study. *Clin Chem* 36:1741-6.

Highsmith, W.E., Friedman, K.J. 1993. The molecular pathology of cystic fibrosis: A clinical laboratory perspective. In Farkas DH (ed.) *Molecular Biology and Pathology.* Academic Press, New York, NY. pp. 159-86.

Highsmith, W.E. 1993. Carrier screening for cystic fibrosis. *Clin. Chem.* 39:706-7.

Highsmith, W.E., Burch, L.H., Zhou, Z., *et al.* A novel cystic fibrosis gene mutation is common in patients with normal sweat chloride concentrations. *New Eng J Med.* (in press).

Kartner, N., Augustinas, O., Jensen, T.J., Naismith, L., Riordan, J.R. 1992. Mislocalization of ΔF508 CFTR in the cystic fibrosis sweat gland. *Nature Genet* 1:321-7.

Kerem, B.S., Rommens, J.M., Buchanan, J.A., *et al.* 1989. Identification of the cystic fibrosis gene: genetic analysis. *Science* 245:1073-80.

Kerem, B.S., Zielenski, J., Markiewicz, D., *et al.* 1990. Identification of mutations in regions corresponding to the two putative nucleotide (ATP)-binding folds of the cystic fibrosis gene. *Proc Nat Acad Sci USA* 87:8447-51.

Kerem, E., Corey, M., Kerem, B-S., *et al.* 1990. The relationship between genotype and phenotype in cystic fibrosis—Analysis of the most common mutation (ΔF508). *New Eng J Med* 323:151-29.

Knowles, M.R., Gatzy, J., Boucher, R.C. 1983. Relative ion permeability of normal and cystic fibrosis nasal epithelium. *J Clin Invest* 71:1410-7.

Knowles, M.R., Gatzy, J., Boucher, R.C. 1981. Increased bioelectric potential difference across respiratory epithelia in cystic fibrosis. *New Eng J Med* 305:1489-95.

Lemna, W.K., Feldman, G.L., Kerem, B.S., et al. 1990. Mutation analysis for heterozygote detection and prenatal diagnosis of cystic fibrosis. *N Engl J Med* 322:291-6.

Myers, R.M., Maniatis, T., Lerman, L.S. 1987. Detection and localization of single base changes by denaturing gradient gel electrophoresis. *Methods Enzymol* 155:501-27.

Osborne, L., Knight, R.A., Santis, G., Hodson, M. 1991. A mutation in the second nucleotide binding fold of the cystic fibrosis gene. *Am J Hum Gen* 47:608-12.

Osborne, L., Santis, G., Schwartz, M., et al. 1992. Incidence and expression of the N1303K mutation of the cytic fibrosis (CFTR) gene. *Hum Gen* 89:653-8.

Orita, M., Iwahana, H., Kanazawa, H., Hayashi, K., Sekiya, T. 1989. Detection of polymorphisms of human DNA by electrophoresis as single-stranded conformation polymorphisms. *Proc Nat Acad Sci USA* 86:2766-70.

Riordan, J.R., Rommens, J.M., Kerem, B.S., et al. 1989. Identification of the cystic fibrosis gene: cloning and characterization of complementary DNA. *Science* 245:1066-73.

Rommens, J.M., Iannuzzi, M.C., Kerem, B.S., et al. 1989. Identification of the cystic fibrosis gene: chromosome walking and jumping. *Science* 245:1059-65.

Sandhoff, K., Conzelmann, E., Neufeld, E.F., Kaback, M.M., Suzuki, K. 1975. The GM2 Gangliosidoses. In Scriver *et al. op cit.*

Sarsfield, J.K., Davies, J.M. Negative sweat tests and cystic fibrosis. *Arch Dis Child* 50:463-6.

Scriver, C.R., Bardanis, M., Cartier, L., Clow, C.L., Lancaster, G.A., Ostrowsky, J.T. 1984. Beta-thalassemia disease prevention: genetic medicine applied. *Am J Hum Gen* 36:1024-9.

Shoshani, T., Augarten, A., Gazit, E., *et al.* 1992. Association of a nonsense mutation (W1282X), the most common mutation in the Ashkenazi Jewish cystic fibrosis patients in Israel, with presentation of severe disease. *Am J Hum Gen* 50:222-8.

Shuber, A.P., Skoletsky, J., Stern, R., Handelin, B.L. 1993. Efficient 12-mutation testing in the CFTR gene: A general model for complex mutation analysis. *Hum Molec Genet* 2:153-8.

Stern, R.C., Boat, T.E., Abramowsy, C.R., Mathews, L.W., Wood, R.E., Doershuk, C.F. 1978. Intermediate-range sweat chloride concentration and Pseudomonas bronchitis: A cystic fibrosis variant with preservation of pancreatic exocrine function. *JAMA* 69:643-6.

Wainwright, B.J., Scambler, P.J., Schmidtke, J., *et al.* 1985. Localization of cystic fibrosis to locus to human chromosome 7cen-q22. *Nature (London)* 318:384-5.

Walker, J.E., Saraste, M., Runswick, M.J., Gay, N.J. 1982. Distantly related sequences in the alpha and beta subunits of ATP synthase, myosin, kinases, and other ATP-requiring enzymes and a common nucleotide binding fold. *EMBO J* 1:945-51.

Waltner, W.E., Boucher, R.C., Gatzy, J.T., Knowles, M.R. 1987. Pharmacotherapy of airway disease in cysic fibrosis. *TIPS* 8:316-20.

Welsh, M.J., Liedtke, C.M. 1986. Chloride and potassium channels in cystic fibrosis airway epithelia. *Nature* 322:467-70.

White, R., Woodward, S., Leppert, *et al.* 1985. A closely linked genetic marker for cystic fibrosis. *Nature (London)* 318:382-4.

Vidaud, M., Fanen, P., Martin, J., Ghanem, N., Nicolas, S., Goossens, M. 1990. Three mutations in the CFTR gene in French cystic fibrosis patients: identification by denaturing gradient gel electrophoresis. *Hum Gen* 85:446-9.

Zielenski, J., Rozmahl, R., Bozon, D., *et al.* 1991. Genomic DNA sequence of the cystic fibrosis transmembrane conductance regulator (CFTR) gene. *Genomics* 10:214-28.

CHAPTER 7

The Molecular Pathology of Duchenne and Becker Muscular Dystrophy

Kenneth J. Friedman

Introduction

Duchenne muscular dystrophy (DMD) is the most common form of inherited muscle-wasting disorder, having an incidence of 1 in 3500 live male births (Emery 1987). Frequency alone is a sufficient imperative for offering diagnostic services to families at elevated risk for this disease. The severity of DMD further underscores the urgency of providing testing. Individuals with DMD experience a slow, debilitating course of disease over two decades that is agonizing for both patients and their families. No therapy or cure is currently available. The earliest symptom in affected boys is the onset of proximal muscle weakness at approximately three years of age. As skeletal muscle degenerates it is invaded by adipose and connective tissue. Calf muscles, in particular, assume a false muscular appearance known as pseudohypertrophy. Difficulty in walking is progressive and culminates in the need for a wheelchair by the early teens. Muscle mass and function continue to be lost, ultimately leading to impaired respiratory function. Death occurs in the late teens or early twenties.

Some individuals exhibit a course of disease similar to DMD but with milder symptoms and a slower disease progression. These individuals suffer from a milder form, Becker muscular dystrophy (BMD). BMD and DMD are allelic, with BMD having a lower incidence of 1 in 30,000. Some patients with BMD father children and retain significant muscle function well past that seen in patients with DMD. One individual has been documented with only moderate impairment in his 60s (England 1990).

DMD and BMD are inherited as X-linked recessive traits; carrier females have a 50% risk of bearing affected sons. Rare instances exist of females expressing DMD-like symptoms. Periodically, errors will occur during meiosis in which dissimilar chromosomes will exchange portions of their structure. These events are called translocations. In nearly every case the cause of a DMD phenotype in females is the result of an X;autosome translocation with a breakpoint impacting the DMD gene (Boyd 1986) accompanied by a preferential inactivation of the unaltered X chromosome.

Molecular diagnostic laboratories have made significant strides in the last five years in providing tests to support clinical diagnoses as well as in providing accurate carrier determination and prenatal testing.

Molecular Pathology of DMD and BMD

In 1959 it was discovered that DMD patients had age-dependent elevations of serum creatine kinase (CK) levels, on the order of 5000 U/L in early childhood compared with a normal range of <200 (Ebashi 1959). Testing serum CK levels therefore provided the first powerful laboratory means of supporting a diagnosis of DMD. Intermediate levels were seen in Becker children. Additionally, an increase in the relative amounts of the muscle-brain (MB) isoenzyme of CK was seen in Duchenne and Becker boys (4–10%) relative to unaffected individuals (0–3%) (Silverman 1976).

The first step in identifying the gene underlying DMD/BMD was its sub-localization on the X chromosome. A study of rare affected females discovered translocation breakpoints on an X chromosome short arm in each woman (Greenstein 1977; Lindenbaum 1979). In 1982 a small boy was described who presented with four different X-linked disorders: DMD, retinitis pigmentosa, McLeod Syndrome and chronic granulomatous disease. Working on the assumption that one large, cytogenetically detectable abnormality was more likely than four independent lesions, karyotype analysis was performed. A large deletion localized to the Xp21 region on the short arm was discovered in this patient (Francke 1985). An early reward of the ensuing race to clone the gene was the discovery by several different laboratories of polymorphic markers linked to the DMD locus. Linkage analysis employing these markers provided the first advance in DMD carrier testing over serum CK testing (Davies 1983).

In 1987 the gene underlying DMD was cloned (Koenig 1987). The protein coding portion of the gene in muscle is now known to be divided into 79 disconnected segments, or exons, dispersed over a 2.5 million base pair region of the X chromosome. Transcription of the gene and processing of the messenger RNA is a complex matter involving multiple initiation sites, several isoforms, and expression in a variety of tissues (Ahn 1993). In skeletal muscle the gene produces a messenger RNA transcript 14,000 base pairs in length that directs the synthesis of dystrophin, a 427,000 MW protein that is rod-like in shape and cytoskeletal in nature (Hoffman 1987; Koenig 1990). Dystrophin represents a small fraction of total muscle protein and has discrete domains bearing homology to the actin-binding regions of α-actinin and the membrane structural protein spectrin.

Immunohistochemical staining of normal muscle tissue cross-sections with anti-dystrophin antibodies demonstrate the protein's association with the inner surface of the sarcolemmal membrane. Although dystrophin is implicated in calcium ion flow across the sarcolemma (Fong 1990), no true catalytic or enzymatic activity has been demonstrated. Recent studies suggest that dystrophin and the glycoprotein complex associated with it (Ervasti 1991) serve to anchor muscle fibers to the extracellular matrix and to maintain the structural integrity of myotubes continually subjected to the stress of contraction (Ohlendieck 1993; Petrof 1993). Mutations within the genes for the 43kd and 50 kd dystrophin-associated glycoproteins result in the autosomally inherited Fukuyama-type congenital musculag dystrophy and severe childhood autosomal-recessive muscular dystrophy, respectively (Matusmura 1994).

Mutations within the dystrophin gene result in the absence of normal dystrophin association with the sarcolemmal membrane. The most prevalent class of mutations underlying both DMD and BMD are, by far, large deletions of the gene. Deletion size can vary greatly, from thousands to a few million base pairs of genomic sequence. Furthermore, deletions may occur at many different locations within the gene (Figure 7.1). The gene's midpoint is frequently implicated in deletion mutations, while a lesser "hot spot" near the 5' end of the gene has been described as well (Koenig 1989; den Dunnen 1989). This allelic heterogeneity has complicated molecular testing for DMD to the extent that a large expanse of sequence must be analyzed in the search for deletions. This is in contrast to a disease like sickle-cell anemia in which every affected individual bears the same mutation, making the screening process rapid and straightforward.

Paradoxically, BMD patients have dystrophin gene deletions which appear much like those underlying Duchenne. Becker patients have been

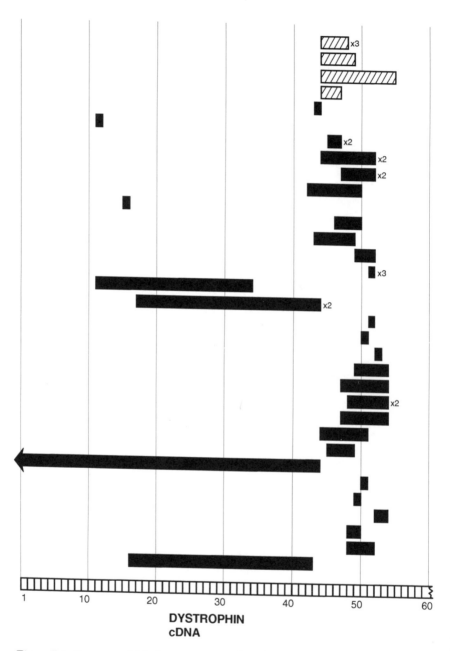

Figure 7.1. Spectrum of deletion mutations within the dystrophin gene. Extent of the dystrophin cDNA is indicated at bottom with exon numbers indicated. Dark bars represent DMD patients seen at the University of North Carolina's Molecular Genetics Laboratory. Hatched bars are BMD patients. Specific deletion mutations encountered in more than one patient are indicated.

reported with large portions of the gene missing (England 1990), while it is not uncommon to find Duchenne individuals lacking single exons. Furthermore, no one region of the gene can be correlated with disease severity. Closer analysis has determined that deletions which disrupt the reading frame of the DNA correlate well with the Duchenne phenotype (Koenig 1989; den Dunnen 1989). Expression of such dystrophin genes results in severely truncated dystrophin molecules that fail to attach to the sarcolemma. If, however, a deletion foreshortens the gene but leaves the reading frame intact, a malformed but semi-functional protein may result, leading to the milder Becker phenotype. Approximately 93% of DMD and BMD patients fit this model. Very large deletions, although technically in-frame, often leave insufficient sequence to code for a viable protein. Alternate start codons, ribosomal frame-shifting and defective mRNA splicing (Winnard 1993) can also subvert the reading frame model.

Approximately 65% of DMD and BMD males have large deletions within the gene and 6–8% carry partial gene duplications (den Dunnen 1989; Angelini 1990). Many of the remaining affected individuals have other sequence anomalies, such as point mutations (Roberts 1992; Prior 1993a), small deletions (Matsuo 1990; Prior 1993b) and splice site defects (Winnard 1993; Wilton 1993). Whereas large deletions can be detected in a number of straightforward ways, detection of point mutations is laborious and not conducive to routine clinical service at this time.

Diagnostic Testing by Mutation Detection

The tissue of choice for DNA analysis is peripheral blood lymphocytes because of the relative non-invasiveness of venipuncture and the likelihood of acquiring ample DNA for testing. Blood is typically collected into vacutainers containing ACD (Acid Citrate-Dextrose) or EDTA as anti-coagulants. Heparin is discouraged as an anti-coagulant because of its potent inhibition of the polymerase chain reaction (PCR), upon which much of the laboratory protocol is based (Beutler 1990).

PCR is a site-specific, enzymatically mediated method for the amplification of targeted regions of the genome (Saiki 1985). For DMD/BMD, the gene segments to be selectively analyzed are the exons of the dystrophin gene. If a deletion has eliminated one or both primer annealing sites, no amplification of that exon is possible, indicating the presence of a deletion.

In order to efficiently detect deletion mutations within the dystrophin gene, it is important to target those exons of the gene frequently absent in DMD and BMD patients. Various protocols have been reported that multiplex, or combine, analyses for 10–12 different dystrophin exons within a single PCR panel (Chamberlain 1990; Beggs 1990; Abbs 1991). One panel claims 98% sensitivity in its ability to detect dystrophin gene deletions. At the UNC Hospitals Molecular Genetics Laboratory, a 22-exon panel has been assembled. These exons are amplified in five multiplexed reactions optimized for successful amplification and compatibility in the electrophoretic separation of the resulting fragments and is described in Table 7.1.

After amplification PCR products are electrophoresed in agarose (3% NuSieve, 1% SeaKem, FMC Bioproducts, Rockland, ME) formulated for resolving small DNA fragments. The gel is stained with ethidium bromide and the bands are visualized by UV illumination. Figure 7.2 demonstrates a multiplex reaction in which panels A, C and E detect the absence of bands corresponding to exons 17, 19, 42 and 43. Appropriate amplification of all other exons, especially exons 16 and 44, determines that this deletion encompasses exons 17–43.

Appropriate controls are essential to the accurate performance of this test. I strongly recommend that single exons are not amplified alone. Absence of a PCR product might be due to failure of amplification rather than a deletion. Running a normal control sample in each panel ensures against mistypings resulting from the failure of an exon to amplify for technical reasons. If a particular exon fails to amplify in both patient and normal control, no deletion has been proven and the test must be repeated. Additionally, a control sample containing water instead of DNA will control for false amplification resulting from reagent contamination.

Detection of a deletion supports a diagnosis of either DMD or BMD. When there is no discernible difference between normal and patient DNA, it is not possible to exclude the diagnosis of Duchenne or Becker muscular dystrophy since the sensitivity of this protocol is approximately 65%. There is no cost-effective means of amplifying all 79 dystrophin gene exons; consequently, uncommon deletions can be missed. Additionally, males bearing duplications, point mutations or other small sequence aberrations will not be detected by the strategy described above.

PCR-based deletion screening has the advantages of speed and technical facility. A more thorough, but laborious, approach to deletion screening is based upon Southern blot methodology (Darras 1988; Prior 1990a). The technique is described in chapter 2. Southern analysis within the

Table 7.1. Dystrophin exons amplified in multiplex reactions for deletion detection.

Multiplex reaction				
A	B	C	D	E
Exon [Size of PCR product (bp)]	Exon [Size of PCR product (bp)]	Exon [Size of PCR product (bp)]	Exon [Size of PCR product (bp)]	Exon [Size of PCR product (bp)]
45 [547]	1 [535]	48 [506]	19 [459]	3 [410]
49 [439]	17 [416]	51 [388]	8 [360]	44 [268]
43 [357]	12 [331]	54 [329]	55 [303]	42 [155]
16 [290]	50 [271]	6 [202]	13 [238]	
47 [181]	52 [113]	53 [100]		

Table 7.1. The 22-exon deletion screening panel used at the UNC Hospitals Molecular Genetics Laboratory. The 22 exons are amplified in five separate multiplex reactions, A–E.

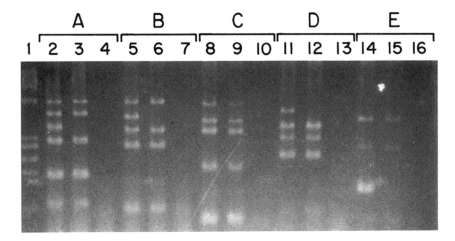

Figure 7.2. Example of screening a patient for dystrophin deletions using a 22-exon PCR panel. PCR primers flanking the exons listed in Table 7.1 were used to amplify the patient's DNA alongside normal control DNA and water blanks. PCR products from each multiplex reaction were electrophoresed in a 4% agarose gel. The ethidium bromide stained gel is shown. Lane 1: φX174 size marker; lanes 2, 5, 8, 11 and 14: normal control DNA; lanes 3, 6, 9, 12 and 15: patient DNA; lanes 4, 7, 10, 13 and 16: water blanks. See Table 7.1 to identify exons based on size. The patient's dystrophin gene lacks exons 17, 19, 42, and 43, yet retains exons 16 and 44; therefore the patient is determined to be deleted for exons 17–43. This is diagnostic of DMD because this deletion disrupts the reading frame of the dystrophin gene.

context of DMD employs DNA probes derived from the coding portion of the dystrophin gene. When the gene was cloned in 1987, a DNA copy of the dystrophin mRNA was made using a viral reverse transcriptase. The resulting cDNA was then cleaved into six fragments which have been made widely available as DNA probes. Once radio-labeled with ^{32}P, these probes will bind specifically to portions of the dystrophin gene which have been restriction-digested, electrophoresed and fixed to a nylon support membrane. Each probe screens a discrete portion of the gene and will, upon autoradiography, be seen as a reproducible pattern of bands on X-ray film representing genomic fragments containing exons of the gene. Absent bands indicate the presence of a deletion. Figure 7.3 depicts several patients screened with the cDNA 8 probe, illustrating the heterogeneous nature of dystrophin gene deletions.

The Southern blot approach has the advantage over PCR of permitting the entire gene to be scanned for deletions; common and uncommon deletions are picked up with equal facility. Partial gene duplications are detected as "junction fragments," bands of altered electrophoretic mobility, on Southern blots (Figure 7.4). Periodically, PCR panels will detect a deletion but fail to map its endpoints. While not critical to making the diagnosis or to providing carrier testing, deletion endpoints are essential to discerning the integrity of the gene's reading frame. Southern analysis can map most endpoints easily and clarify a prognosis of either the severe Duchenne course of disease or the milder Becker phenotype.

Southern analysis has its disadvantages, too. The process of screening the entire dystrophin gene requires sequential hybridization with 6 DNA probes derived from the dystrophin cDNA, each of which can take one week to complete. Southern analysis currently requires the use of radioisotopes. Whereas several colorimetric and chemiluminescent protocols are now available, difficulties resulting from low sensitivity have limited their adoption into routine service. The need for detecting the few duplications and rare deletions should be weighed against the volume of patients seen and the effort involved. Lastly, Southern analysis in no way aids mutation detection in the 30% of patients lacking large deletions or duplications.

When a laboratory is choosing which methodology to apply to mutation detection, joint application of both PCR- and Southern-based methodologies should be considered as a way to mitigate the deficiencies of both. Rapid application of a PCR multiplex panel to newly arrived samples will

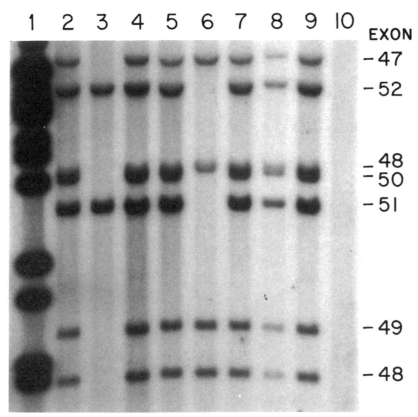

Figure 7.3. Example of an autoradiogram after Southern-based analysis of seven patients' DNA. The HindIII-digested DNA samples were electrophoresed in 1% agarose to separate DNA fragments on a size-dependent basis and then transferred to a nylon membrane for hybridization with the DMD probe cDNA 8. Lane 1: BstEII-digested λ size marker; lane 2: normal control; lanes 3–10: patient DNA samples. The patient in lane 3 is missing exons 47, 48, 49, and 50. The patient in lane 6 lacks exons 50, 51, and 52. The patient in lane 10 lacks every exon (47–52) detected by hybridization with the probe cDNA 8.

detect the great majority of deletions immediately, and Southern blot hybridization can be used to map the deletion endpoints where necessary. In this manner, a smaller number of patients without PCR-detectable deletions would be subjected to an extended Southern-based panel. Lastly, confirming molecular results by two complementary approaches can only enhance the integrity of a laboratory's output.

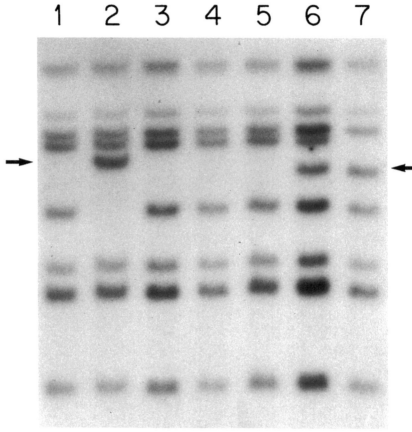

Figure 7.4. Example of junction fragment analysis. Southern blot hybridization of HindIII-digested patient DNA samples was performed as described in Figure 7.3, except hybridization was performed with the DMD probe cDNA 2B-3. Different junction fragments (indicated with arrows) observed in patients in lanes 2 and 7. Lane 6: DNA from the mother of individual in lane 7 bears both the normal band and the junction fragment seen in her son, thus establishing carrier status.

Carrier Detection Strategies

Carrier testing is most commonly sought by women of reproductive age who have family histories of Duchenne or Becker muscular dystrophy. Before the advent of molecular diagnostic techniques, a physician's ability to make accurate determinations of carrier status was limited to determining *a priori* risks based upon family history. However, determination of an individual's *a priori* carrier risk is complicated from several perspectives. Most critical is the high rate of new mutations that cause disease. Statistically, one-third of all individuals affected with X-linked lethal disorders are expected to be the result of novel, or sporadic, mutations (Haldane 1935). Expressed another way, only two-thirds of mothers of DMD or BMD boys are at 50% risk of bearing another son with this disease. Most of the other mothers have no greater risk than the general population since none of their remaining oocytes are likely to bear the mutation.

One's ability to determine carrier status improves if one obtains an accurate pedigree of the family that establishes prior incidence of the disease in the family. Since it is highly unlikely that two novel dystrophin gene mutations would occur independently within one extended family, all mothers and the common grandmother of two or more affected individuals are deemed to be obligate carriers. Other females would have an *a priori* risk derived from their relationship to the affected individual.

Periodically a potential carrier will herself present to a neurology clinic with muscle weakness. These women are regarded as manifesting carriers (Jacobs 1981). During female development, each cell has one of its two X chromosomes inactivated. This process is known as lyonization and is random from cell to cell. If, by chance, a DMD or BMD carrier has a disproportionate number of her normal X chromosomes inactivated in her muscle tissues, she will express some features of this disease. In fact, greatly skewed X-inactivation has led to full DMD symptomology in one of two monozygotic twin females (Zneimer 1993). However, since only about 5% of DMD/BMD carrier women manifest muscle weakness, this presentation has limited applicability to carrier testing.

Serum creatine kinase levels, dramatically elevated in affected individuals, can be abnormally high in carrier females as well (>150 U/L). However, the utility of serum CK levels for carrier testing is limited. A high CK level in female relatives of DMD boys is a strong predicter of carrier status, but because 30% of carriers exhibit normal CK levels, normal CK levels do not exclude carrier status. Therefore, a normal serum CK is a poor indicator of non-carrier status (Gruemer 1984). Despite this shortcoming, creatine kinase testing can aid carrier determination. In order to enhance its diagnostic value, I recommend that three serum CK levels be performed on different days to ensure the test's validity.

Molecular diagnostic approaches to carrier testing followed the discovery and cloning of the dystrophin gene in 1987. The approaches were based on the expectation that if a patient has a deletion mutation within the gene, any carrier females in his family should possess one X chromosome that is identically deleted. Carrier females would, therefore, be hemizygous for, (i.e., have one copy only of) that portion of the gene absent in their affected relative. This reduced dosage was first analyzed by Southern blot hybridization techniques (Prior 1989; Laing 1989). The autoradiogram in Figure 7.5A illustrates the patterns seen for an individual with a deletion, his mother, his sister and a normal control. Those bands absent in the affected male should be present with decreased intensity in any female carrying the same deletion. Quantitative analysis is accomplished through densitometric scanning of the film (Figure 7.5B), and by comparing the ratio of the top two bands (one of which the affected boy lacks) in the putative carriers with that of the normal control. In this instance, positive carrier status for both mother and daughter has been established.

Approximately 5% of affected individuals generate junction fragments of altered mobility upon Southern analysis. Junction fragments are the result of partial gene duplications or represent the direct detection of deletion breakpoints, and make carrier testing simple when detected in potential carriers. As illustrated in Figure 7.4, the junction fragment seen in DNA from a DMD boy (lane 7) is present in DNA from his mother (lane 6), who also possesses the normal band. Dosage analysis is supplanted by the simple presence or absence of the aberrant band.

Although uncommon, junction fragments have uncovered another potential source of error impinging upon carrier testing. Laboratory analysis is performed using DNA derived from peripheral blood lymphocytes under the assumption that somatic cell DNA is effectively identical to that of the gametes. In some cases this assumption is incorrect; germline mosaicism may occur as a result of the developmental divergence of the genetic make-ups of somatic and germ cells. Should the dystrophin gene in a woman's germ line bear a mutation not present in her somatic DNA, an incorrect diagnosis of non-carrier status can be made (Bakker 1989). Analysis of oocyte DNA is not routinely feasible, so the only reported examples of germline mosaicism arise from junction fragment analyses (Prior 1992). In these cases affected boys generate junction fragments absent from their mothers' peripheral blood DNA. Sisters that carry the altered band, however, are deemed carriers, suggesting an elevated carrier risk for their mothers. Germline mosaicism can also be documented by prenatal diagnosis and dosage analysis.

The recurrence risk for DMD/BMD due to germline mosaicism for a

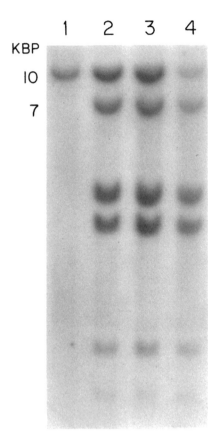

Figure 7.5A. (left) Example of carrier determination by Southern blot hybridization and dosage analysis. An autoradiogram is shown of HindIII-digested DNA probed with DMD cDNA 8. Lane 1: proband exhibiting a deletion of exons 48–52; lane 2: mother; lane 3: sister; lane 4: normal control. The 10 kb fragment contains exon 47 and the 7 kb band contains exon 52. True carriers of the proband's deletion should exhibit twice the intensity of the 10 kb band relative to the 7 kb band.

Figure 7.5B. (below) Densitometric scan of autoradiogram in Figure 7.5A to determine carrier status. Scan 1: proband; scan 2: mother; scan 3: sister; scan 4: normal control. Autoradiogram was analyzed using a Shimadzu CS 9000U scanning densitometer. Both mother and sister of the proband are determined to be carriers of the familial deletion since the intensity of their 10 kb bands is twice that of their 7 kb bands relative to the normal control.

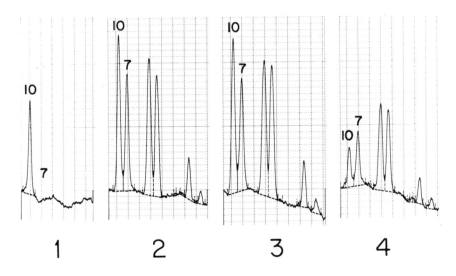

proven novel mutation has been estimated at 8% (Bakker 1989). Furthermore, it has been reported that new mutations in the proximal portion of the gene have a higher recurrence risk than new mutations in the midregion (Passos-Bueno 1992). Not all laboratories have made this observation, however. A laboratory's best approach to detecting incidents of germline mosaicism resides in testing sisters of affected boys for their brothers' mutations. Detection of aberrant patterns in a sister by junction fragment or dosage analysis will establish germline mosaicism. A negative result for a sister will not disprove mosaicism, since it is impossible to assess how many of the mother's oocytes bear the mutation.

As described earlier in the section on mutation detection, each Southern blot hybridization is a cumbersome procedure requiring a week of patient manual effort. PCR has demonstrated its utility for mutation detection and can be modified to address the needs of carrier testing as well (Prior 1990b; Ioannou 1990). A typical PCR reaction requires approximately 30 cycles of amplification to allow adequate visualization of DNA products. By this time, however, one has long since exceeded the linear phase of the reaction, making it impossible to quantitate the bands and extrapolate to the original dosage. A smaller number of amplification cycles will limit the reaction to the linear phase but insufficient material will have been amplified by this time to be seen by ethidium bromide staining, let alone to be quantitated.

We have developed a variation of published procedures which involves the inclusion of ^{32}P-dCTP in the PCR reaction to permit carrier determination. Two exons, one deleted in the patient, the other intact, are co-amplified for the affected individual, his female relatives and normal controls. Isotope is incorporated into the amplified material. After 18–20 amplification cycles the PCR product is concentrated and unincorporated isotope is removed via centrifugation (Centricon-100, Amicon, Beverly, MA). After 2 hours of electrophoresis in a 50 ml, 2% SeaKem agarose gel, the gel is dried onto filter paper and exposed to X-ray film for 10–30 minutes. Short exposure times help ensure that the linearity of the X-ray film is not exceeded. After autoradiography densitometric scanning is used to compare the dosage of the two amplified exons in the potential carriers against the normal controls.

An example of our isotopic PCR method for carrier determination is shown in Figure 7.6A. The affected boy is deleted for exon 6. Exon 48

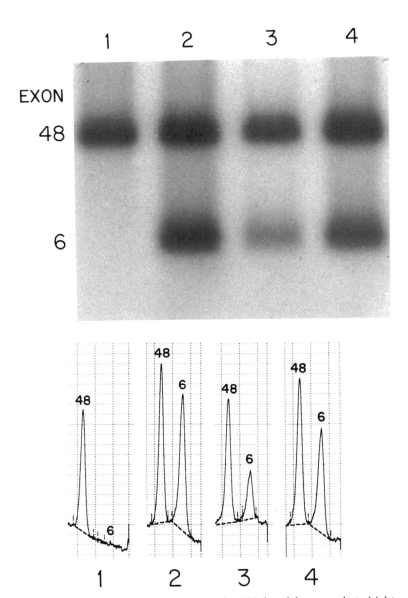

Figure 7.6A. Example of carrier determination by PCR-based dosage analysis. Multiplex, short-cycle, isotopic PCR was performed for the dystrophin gene exons 6 and 48 using DNA from an affected male, his sister, his mother, and a normal control. After amplification the PCR products are electrophoresed in an agarose gel. The gel is then dried onto filter paper and autoradiography was performed. Lane 1: proband (lacking exon 6); lane 2: sister; lane 3: mother; lane 4: normal control.

Figure 7.6B. Densitometric scan of autoradiogram in Figure 7.6A. Scan 1: proband; scan 2: sister; scan 3: mother; scan 4: normal control. Mother has a reduced dosage of exon 6, establishing her carrier status. The sister's dosage is identical to that seen in the normal control; therefore, she is not a carrier.

Table 7.2. PCR-based polymorphisms used to diagnose Duchenne and Becker muscular dystrophy.

Locus Name	Location	Class	Enzyme	Reference
None specified	5' end of gene	RFLP	MaeIII	Roberts 1991
5' DYS II	Near brain promoter	Dinucleotide polymorphism	NA	Feener 1991
5' DYS MSA	Near muscle promoter	Dinucleotide polymorphism	NA	Oudet 1991
p87-8	cDNA 2B-3 region	RFLP	TaqI	Roberts 1989
p87-15	cDNA 2B-3 region	RFLP	BamHI	Roberts 1989
p87-15	cDNA 2B-3 region	RFLP	XmnI	Roberts 1989
STR-44	Intron 44	Dinucleotide polymorphism	NA	Clemens 1991
STR-45	Intron 45	Dinucleotide polymorphism	NA	Clemens 1991
None specified	Intron 46	Dinucleotide polymorphism	NA	Powell 1991
None specified	Exon 48	RFLP	MseI	Yau 1991
STR-49	Intron 49	Dinucleotide polymorphism	NA	Clemens 1991
STR-50	Intron 50	Dinucleotide polymorphism	NA	Clemens 1991
None specified	3' untranslated	Dinucleotide	NA	Oudet 1991

NA: not applicable.

was amplified as an internal control. The band intensity ratio for exons 48 and 6 was determined by densitometric scanning for the boy, his sister, his mother and a normal control. The results are shown in Figure 7.6B. In this example, the mother was found to carry one-half the normal dosage of exon 6 relative to that seen in the normal control and is, therefore, a carrier. Her daughter, however, was fortunate to have received her mother's undeleted X chromosome and was, therefore, not a carrier.

Dosage techniques are the method of choice for determinations of carrier status. Whereas radioisotopes are necessary to the protocol described above, non-isotopic approaches are appearing in the literature (Schwartz 1992). Applied Biosystems (Culver City, CA) has manufactured the Fluorescent Fragment Analyzer, a largely automated, high-resolution apparatus for the resolution and quantitation of numerous DNA fragments. When amplified with fluorescently tagged primers, many dystrophin exons can be compared for relative intensity with great precision. Although the cost is, perhaps, prohibitively expensive for the average laboratory, this device is an indication that a new generation of lab instrumentation is on the way (see chapter 2).

The 30% of males affected with DMD or BMD who do not bear detectable deletions within their dystrophin genes typically have point mutations or very small deletions. These small sequence variations appear to be unique to each affected male and are scattered across the 14,000 base pairs of the gene's coding region. Their detection is not trivial. Carrier testing based upon point mutations has been reported (Yau 1993), but testing for them is a family-specific matter that is usually beyond the scope of a routine service laboratory.

The female relatives of the males with DMD or BMD without detectable deletions are, therefore, placed at a disadvantage. Without a mutation to test for in any directed manner, the laboratory must resort to indirect, linkage-based approaches. Linkage analysis is predicated upon the co-inheritance of closely apposed genetic loci (see chapter 5). The likelihood that any two genetic loci on the X chromosome will be co-inherited by a son from his mother is inversely proportional to the distance between them. The further apart the two loci are, the more likely a meiotic recombination will occur between them. To date, there are several polymorphic loci within the dystrophin gene that may be analyzed by PCR in a clinical setting. These are shown in Table 7.2. These polymorphisms are intragenic, i.e., they reside inside the gene. They have greater diagnostic value than extragenic markers, located outside the gene, because of the reduced risk of a recombination occurring between the polymorphism and the site of the mutation. Many more RFLPs are available for analysis by Southern blotting. For genetic disorders arising from genes smaller than the dystrophin gene, the existence of intragenic markers would effectively elim-inate the recombinational risk of error. However, the enormous size of the dystrophin gene and the presence of a hot spot for recombination within it are complicating factors. It has been reported that linkage analysis pred-icated upon a single intragenic polymorphism in the dystrophin gene has an inherent error rate of 12% (Abbs 1990). Using multiple informative markers can reduce this source of error. For example, two informative markers, one at either end of the gene, will preclude all but the risk of a dou-ble recombination within the gene's 2.5 million base pairs. This risk is estimated to be 12% of 12%, or approximately 1.4%.

Polymorphisms in and around the dystrophin gene fall into two general categories: restriction fragment length polymorphisms (RFLPs) and microsatellite markers. RFLPs typically have two alleles defined by the presence or absence of a recognition sequence for a restriction endonuclease. After amplification of a region bracketing a polymorphism, the product is digested with an appropriate restriction enzyme and electrophoresed to resolve the alleles by size. By convention the larger, uncut allele is desig-nated "1" and the smaller, cleaved fragment(s), "2." Analysis of RFLPs is easy and straightforward, but their limited number of alleles reduces each marker's informativity, since critical females in a DMD pedigree must be heterozygous at polymorphic loci in order to distinguish their X chromo-somes.

A more informative class of polymorphisms known as microsatellites has been described in recent years and they have demonstrated their utility for DMD/BMD linkage analysis. Microsatellite markers are characterized by a variable number of tandem repeats of a short DNA sequence, the total length of which varies from one individual to the next. These markers often have more than ten alleles. The greater the number of alleles, the less likely that an individual is homozygous at that locus. The short tandem repeat (STR) markers listed in Table 7.2 are dinucleotide polymorphisms that have enough alleles that the chances of encountering a true homozy-gous female are small. When a homozygous female pattern is encountered in a DMD/BMD family, it is strong evidence that she may be, in actuality, hemizygous for this locus due to a deletion on one X chromosome and might, therefore, be a carrier (Clemens 1991). This approach to carrier test-ing may have applicability to cases when the affected individual is deceased or his DNA is otherwise unattainable.

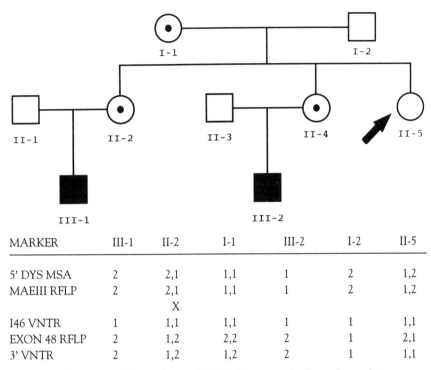

MARKER	III-1	II-2	I-1	III-2	I-2	II-5
5' DYS MSA	2	2,1	1,1	1	2	1,2
MAEIII RFLP	2	2,1	1,1	1	2	1,2
		X				
146 VNTR	1	1,1	1,1	1	1	1,1
EXON 48 RFLP	2	1,2	2,2	2	1	2,1
3' VNTR	2	1,2	1,2	2	1	1,1

Figure 7.7. Pedigree of a familial case of DMD. Genotypes for five polymorphisms are shown beneath the pedigree. A recombination (indicated by the "X") has occurred between the MaeIII and Mse1 polymorphisms. The two affected males share the gene portion proximal to the Mse1 marker; therefore, the familial mutation must reside in this region. The consultand indicated by the arrow was determined not to have received that portion of the dystrophin gene implicated in disease.

A pedigree representing a familial case of DMD is provided in Figure 7.7. In the family shown the two affected individuals, III-1 & III-2, have no detectable deletions. Obligate carrier status is easily established for individuals I-1, II-2, and II-4, but it was II-5 who sought carrier testing. A linkage study using five different polymorphic markers determined that an intragenic recombination has occurred between III-1 and his mother. Knowing the location of the recombinational event allowed us to focus attention on that portion of the dystrophin gene shared by the affected boys, in this case, the 3´ portion of the gene. The consultand was determined to have received the maternal chromosome that did not carry the DMD mutation. Barring a second undetected intragenic recombination, the consultand was, therefore, found not to be a carrier.

Sporadic cases represent additional diagnostic dilemmas in the absence of a detectable deletion. Becker muscular dystrophy is a less straightforward clinical diagnosis than DMD due to its milder progression. An uncertain diagnosis of BMD raises the issue of genetic heterogeneity, or the possibility that similar disease phenotypes can arise from mutations in other genes. Besides other X-linked myopathies, there are rare autosomal disorders which may be confused with Becker if the pedigree does not exclude autosomal inheritance patterns. If a patient's myopathy has an origin other than a mutation within the dystrophin gene, then the application of dystrophin-based linkage analysis is nonsensical at best. It has been estimated that 8–12% of boys diagnosed with DMD may in fact bear autosomal recessive forms of the disease (Vainzof 1991a). It is the clinician's responsibility to provide an accurate neuromuscular evaluation of the patient while the collection of an accurate pedigree falls to the genetic counselor.

With regard to a true sporadic case, linkage analysis alone is of limited value due to the high new mutation rate for this disease. Either the mother acquired a dystrophin gene mutation from her parents, be it familial but unexpressed, or of novel origin, or her affected son is the result of the first mutational event in the pedigree. With no *a priori* knowledge of the mutation's origin, linkage analysis alone is insufficient to evaluate carrier status with any credibility.

Bayesian analysis is a statistical tool by which the various risk factors can be compiled to provide a valid numerical assessment of carrier risk. Figure 7.8 illustrates a pedigree in which there is no family history of DMD or BMD. II-1 has no detectable deletion and serum creatine kinase levels tested normal in both mother and sisters. In this example, linkage analysis with three widely dispersed markers determined that both sisters received the same maternal X chromosome as their affected brother. The results of the Bayesian analysis are shown in Table 7.3. Although this form of analysis lacks the warm glow of certainty associated with dosage-based carrier testing, it represents the best statistical means available for providing risk assessments to sporadic, non-deletion families.

Prenatal Diagnosis

As with any severe genetic disease, families at risk for DMD or BMD often seek the assistance of laboratories and genetic counselors in making informed reproductive decisions. Some families will consider termination

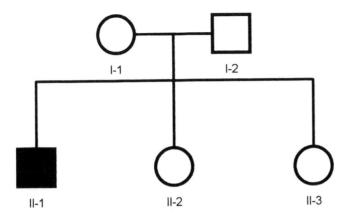

Figure 7.8A. Pedigree of a family in which a sporadic case of DMD was ascertained. II-1 has no detectable deletion. I-1, II-2, and II-3 have normal serum CK levels.

Determination of mother's carrier risk:

	I-2 is a carrier	I-2 is not a carrier
Prior Risk	0.660	0.330
Conditional Risk (Normal CK)	0.300	1.000
Joint Risk (Prior x Conditional)	0.198	0.330
Joint Sum (sum of both Joint risks)		0.528
Posterior Risk (Joint risk / Joint sum)	**0.375**	0.625

Determination of sisters' carrier risk (same for each sister)

	Is a carrier	Is not a carrier
Prior Risk	0.375*	0.625
Conditional Risk (Normal CK)	0.300	1.000
Joint Risk (Prior x Conditional)	0.113	0.625
Joint Sum (sum of both Joint risks)		0.738
Posterior Risk (Joint risk / Joint sum)	0.150	0.850

Figure7.8B. Example of Bayesian analysis to assess carrier risks in the family. Mutation analysis is not possible because no detectable deletion was found in the affected individual. Linkage analysis is not possible for a sporadic case. The result of the analysis is that each sister has a carrier risk of 15%, barring a small recombinational error.

* This number is derived from the mother's carrier risk and from the linkage analysis demonstrating that the sisters each received the same maternal X chromosome as their affected brother.

of a pregnancy if a male fetus is demonstrated to bear a dystrophin gene mutation. Others seek the information to better prepare themselves emotionally for the long-term care of the child once it is born.

Assuming a diagnosis of carrier status, carrier females and their spouses are typically counseled that their risk of bearing an affected son or a carrier daughter is 50%. If prenatal testing is requested then, once a carrier female is pregnant, fetal tissue must be collected by either amniocentesis or chorionic villus sampling (CVS). The former procedure is typically performed at between 15–18 weeks of gestation. Amniocytes are grown in tissue culture for a week to ten days to ensure an ample number of cells for DNA testing. CVS has the dual advantages of being performed earlier (9–11 weeks) and yielding more tissue for immediate testing without the need for culturing. Growing these cells in culture, however, minimizes the risk of residual maternal cell contamination that might mask the presence of a deletion.

Molecular analysis of a potential male DMD fetus represents the test of greatest priority to the diagnostic laboratory. Gender determination is performed either by karyotyping or Y chromosome-specific PCR. Male fetuses are screened for the familial mutation, or linkage analysis is performed as described earlier. Test results are communicated to the genetic counselor who transmits them to the family (see chapter 9).

As mentioned earlier, undetected germline mosaicism is a potential source of error with regard to carrier determinations. When females with DMD/BMD relatives are determined not to be carriers, I strongly recommended that they seek prenatal diagnosis for all future pregnancies.

Fetal muscle biopsy using the procedures described below has been reportedly used for prenatal diagnosis in cases where there is no known familial deletion to detect and when linkage analysis is uninformative (Kuller 1992). Fortunately, linkage analysis using the highly polymorphic microsatellite markers currently available is invariably informative, obviating the need for a difficult and invasive procedure.

The current limitations of linkage testing include issues of genetic heterogeneity, inaccurate family history, the unavailability of key family members, recombinational error, germline mosaicism and the rate of new mutations.

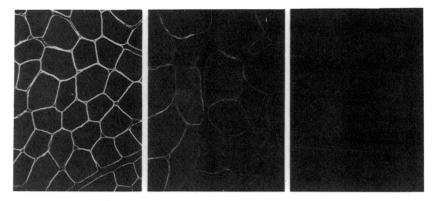

Figure 7.9. Example of immunohistochemical staining for dystrophin in muscle tissue cross-sections. Left: Normal staining. Middle: Patchy staining characteristic of BMD. Right: Negative staining characteristic of DMD. (Courtesy of K. Arahata, National Institute of Neuroscience, Tokyo. Reproduced from Thompson et al. 1991 with permission.

Testing Using Dystrophin Protein Chemistry

As with any genetic disorder, symptoms arise not from the mutations per se, but rather from the dysfunctional gene product. Many laboratories exploit this by incorporating direct analysis for the dystrophin protein into their DMD molecular testing protocols. Western blots can be used to evaluate the abundance as well as the size of the dystrophin molecule in solubilized skeletal muscle tissue. Absence of dystrophin is associated with DMD. If the protein is reduced either in size or abundance, diagnosis of the Becker phenotype is indicated (Hoffman 1989). ELISA-based approaches to protein detection are also available for Western blots (Byers 1992).

Immunohistochemical staining of muscle tissue cross-sections produces recognizable patterns associated with either the normal, Duchenne or Becker phenotype (Hoffman 1988). Duchenne patients fail to show the strong staining seen with normal tissue while the milder Becker phenotype is associated with a patchy, more diffuse pattern. An example is shown in Figure 7.9.

The chief disadvantage of protein-based testing is the requirement for muscle tissue; a punch biopsy is more invasive than a peripheral blood draw. Further, the muscle tissue of advanced cases is largely invaded with adipose and connective tissue that can impair one's ability to acquire a sufficient sample.

The primary utility of protein testing is to clarify a questionable diagnosis in the absence of a dystrophin gene deletion. A physician may seek to confirm or refute an uncertain diagnosis of Becker muscular dystrophy by ordering a deletion screen. If no deletion is found, Western blot or immuno-histochemical staining techniques can assess the integrity of the protein directly. It is important to employ an antibody panel directed toward more than one dystrophin epitope to ensure adequate detection. An aberrant dys-trophin pattern confirms the diagnosis and allows clinical management to proceed with greater confidence. Furthermore, detecting abnormal dys-trophin in an affected male establishes the validity of linkage analysis for carrier testing of the patient's female relatives. An apparently normal dys-trophin staining pattern suggests a different etiology for the patient's dis-ease.

Using direct immunohistochemical screening for dystrophin carrier testing has been an active and controversial field. Manifesting carriers routinely exhibit abnormal dystrophin-staining patterns wherein some muscle fibers seen in cross-section stain positive and others do not (Morandi 1990). However, asymptomatic carriers have been reported with similar mosaic staining (Bernier 1993) or with normal patterns (Vainzof 1991b). In the absence of a consensus it is difficult to exclude carrier status by nor-mal dystrophin immunohistochemistry at this time.

Summary and Future Prospects

Providing molecular diagnostic services for families at risk for Duchenne or Becker muscular dystrophy poses numerous challenges in the endeavor to provide the most accurate information. The complexities of mutation detection, carrier testing, prenatal diagnosis and dystrophin protein chem-istry all demand a grasp of theory, attention to detail, and continual effort to keep abreast of new developments.

Pivotal to the future success of DMD/BMD testing will be the stan-dardization of techniques, automation and quality control. At present, laboratories have developed their own permutations of how best to deliver diagnostic services. In order to establish uniformity of quality assurance, the College of American Pathologists is instituting guidelines to assist labo-ratories in maintaining the highest possible standards of service. Further, the establishment of national or regional proficiency testing will serve similar objectives.

Suggested Reading

Ahn, A.H., Kunkel, L.M. 1993. The structural and functional diversity of dystrophin. *Nature Genet.* 3, 283-91.

Beggs, A. H. Multiple PCR for identification of dystrophin gene deletions in patients with Duchenne and Becker muscular dystrophy. In: *Current Protocols in Human Genetics* (in press).

Prior, T.W. 1993. Duchenne and Becker muscular dystrophy: Current diagnostics. In: *Molecular Biology and Pathology: A Guidebook for Quality Control.* (ed. Farkas, D.H.) 187-200 (Academic Press, San Diego, CA).

References

Abbs, S., Roberts, R.G., Mathew, C.G., Bentley, D.R., Bobrow, M. 1990. Accurate assessment of intragenic recombination frequency within the Duchenne muscular dystrophy gene. *Genomics* 7, 602-6.

Abbs, S., Yan, S.C., Clarke, S., Mathew, C.G., Bobrow, M. 1991. A convenient multiplex PCR system for the detection of dystrophin gene deletions: A comparative analysis with cDNA hybridization show mistypings by both methods. *J Med Genet* 28, 304-11.

Ahn, A.H., Kunkel, L.M. 1993. The structural and functional diversity of dystrophin. *Nature Genet* 3, 283-91.

Angelini, C., Beggs, A.H., Hoffman, E.P., Fanin, M., Kunkel, L.M. 1990. Enormous dystrophin in a patient with Becker muscular dystrophy. *Neurology* 40, 808-12.

Bakker, E., Veenema H., den Dunnen, J.T., *et al.* 1989. Germinal mosaicism increases the recurrence risk for "new" Duchenne muscular dystrophy mutations. *J Med Genet* 26, 553-9.

Beggs, A.H., Koenig, M., Boyce, F.M., Kunkel, L.M. 1990. Detection of 98% of DMD/BMD deletions by polymerase chain reaction. *Hum Genet* 86, 45-8.

Beutler, E., Gelbart, T., Kuhl, W. 1990. Interference of heparin with the polymerase chain reaction. *BioTechniques* 9, 166.

Boyd, Y., Buckle, V.J., Holt, S.M., Munro, E.A., Hunter, D., Craig, I.W. 1986. Muscular dystrophy in girls with X;autosome translocations. *J Med Genet* 23, 484-90.

Byers, T.J., Neumann, P.E., Beggs, A.H., Kunkel, L.M. 1992. ELISA quantitation of dystrophin for the diagnosis of Duchenne and Becker muscular dystrophies. *Neurology* 42, 570-6.

Chamberlain, J.S., Gibbs, R.A., Ranier, J.E., Caskey, C.T. 1990. Multiplex PCR for the diagnosis of Duchenne muscular dystrophy. In: *PCR Protocols: A Guide to Methods and Applications*. (eds. Innis, M., Gelfand, D., Sninski, J., White, T.) 272-81 (Academic Press, Orlando).

Clemens, P.R., Fenwick, R.G., Chamberlain, J.S., *et al.* 1991. Carrier detection and prenatal diagnosis in Duchenne and Becker muscular dystrophy families, using dinucleotide repeat polymorphisms. *Am J Hum Genet.* 49, 951-60.

Darras, B.T., Blattner, P., Harper, J.R., Spiro, A.J., Alter, S., Franke, U. 1988. Intragenic deletions in 21 Duchenne muscular dystrophy (DMD)/Becker muscular dystrophy (BMD) families studied with the dystrophin cDNA: Location of breakpoints on HindIII and BglII exon-containing fragment maps, meiotic and mitotic origin of mutations. *Am J Hum Genet* 43, 620-9.

Davies, K.E., Pearson, P.L., Harper, P.S. 1983. Linkage analysis of two cloned DNA sequences flanking the Duchenne muscular dystrophy locus on the short arm of the human X chromosome. *Nucl Acids Res* 11, 2303-12.

den Dunnen, J.T., Bakker, E., Klein Breteler, E.G., Pearson, P.L., van Ommen, G.J.B. 1989. Topography of the Duchenne muscular dystrophy (DMD) gene: FIGE and cDNA analysis of 194 cases reveals 115 deletions and 13 duplications. *Am J Hum Genet* 45, 835-47.

Ebashi S., Toyokura, Y., Momoi, H., *et al.* 1959. High creatine phosphokinase activity of sera of progressive muscular dystrophy. *J Biochem* 46, 103-4.

Emery, A.E.H. 1988. *Duchenne Muscular Dystrophy* (2nd ed.) 1-317 (Oxford University Press).

England, S.B., Nicholson, L.V.B., Johnson, M.A., *et al.* 1990. Very mild muscular dystrophy associated with the deletion of 46% of dystrophin. *Nature* 343, 180-2.

Ervasti, J.M., Campbell, K.P. 1991. Membrane organization of the dystrophin-glycoprotein complex. *Cell* 66, 1121-31.

Feener, C.A., Boyce, F.M., Kunkel, L.M. 1991. Rapid detection of CA polymorphisms in cloned DNA: Application to the 5' region of the dystrophin gene. *Am J Hum Genet* 48, 621-7.

Fong, P., Turner, P.R., Denetclaw, W.F., Steinhardt, R.A. 1990. Increased activity of calcium leak channels in myotubes of Duchenne human and mdx mouse origin. *Science* 250, 673-6.

Francke, U., Ochs, H.D., de Martinville, B., *et al.* 1985. Minor Xp21 chromosome deletion in a male associated with expression of Duchenne muscular dystrophy, chronic granulomatous disease, retinitis pigmentosa, and McLeod syndrome. *Am J Hum Genet* 37, 250-67.

Greenstein, R.M., Reardon, M.P., Chan, T.S. 1977. An X;autosome translocation in a girl with Duchenne muscular dystrophy (DMD): Evidence for a DMD gene localization. *Pediatr Res* 11, 457A.

Gruemer, H.D., Miller, W.G., Chinchilli, V.M., *et al.* 1984. Are reference limits for serum creatine kinase valid in detection of the carrier state for Duchenne muscular dystrophy? *Clin Chem* 30, 724-30.

Haldane, J.B.S. 1935. The rate of spontaneous mutation of a human gene. *J Genet* 31, 317-26.

Hoffman, E.P., Brown, R.H. Jr., Kunkel, L.M. 1987. Dystrophin: The protein product of the Duchenne muscular dystrophy locus. *Cell* 51, 919-928.

Hoffman, E.P., Fischbeck, K.H., Brown, R.H., *et al.* 1988. Characterization of dystrophin in muscle-biopsy specimens from patients with Duchenne or Becker muscular dystrophy. *N Engl J Med* 318, 1363-8.

Hoffman, E.P., Kunkel, L.M., Angelini, C., Clarke, A., Johnson, M., Harris, J.B. 1989. Improved diagnosis of Becker muscular dystrophy by dystrophin testing. *Neurology* 39, 1011-7.

Ioannou, P., Christopoulos, G., Panayides, K., Kleanthous, M., Middleton, J. 1992. Detection of Duchenne and Becker muscular dystrophy carriers by quantitative multiplex polymerase chain reaction analysis. *Neurology* 42, 1783-90.

Jacobs, P.A., Hunt, P.A., Mayer, M., Bart, R.D. 1981. Duchenne muscular dystrophy (DMD) in a female with an X/autosome translocation: Further evidence that the DMD locus is at Xp21. *Am J Hum Genet* 33, 531-8.

Koenig, M., Hoffman, E.P., Bertelson, C.J., Monaco, A.P., Feener, C., Kunkel, L.M. 1987. Complete cloning of the Duchenne muscular dystrophy (DMD) cDNA and preliminary genomic organization of the DMD gene in normal and affected individuals. *Cell* 50, 509-17.

Koenig, M., Beggs, A.H., Moyer, M., *et al.* 1989. The molecular basis for Duchenne versus Becker muscular dystrophy: Correlation of severity with type of deletion. *Am J Hum Genet* 45, 498-506.

Koenig, M., Kunkel, L.M. 1990. Detailed analysis of the repeat domain of dystrophin reveals four potential hinge segments that may confer flexibility. *J Biol Chem* 265, 4560-6.

Kuller, J.A., Hoffman, E.P., Fries, M.H., Golbus, M.S. 1992. Prenatal diagnosis of Duchenne muscular dystrophy by fetal muscle biopsy. *Hum Genet* 90, 34-40.

Laing, N.G., Siddique, T., Bartlett, R., *et al.* 1989. Duchenne muscular dystrophy: Detection of deletion carriers by spectrophotometric densitometry. *Clin Genet* 35, 393-8.

Lindenbaum, R.H., Clark, G., Patel, C., Moncrieff, M., Hughes, J.T. 1979. Muscular dystrophy in an X;1 translocation female suggests that Duchenne locus is on X chromosome short arm. *J Med Genet* 16, 389-92.

Matsumura, K., Campbell, K.P. 1994. Dystrophin-glycoprotein complex: Its role in the molecular pathogenesis of muscular dystrophies. *Muscle and Nerve* 17:2-15.

Matsuo, M., Masumura, T., Nakajima, T., *et al.* 1990. A very small frame-shifting deletion within exon 19 of the Duchenne muscular dystrophy gene. *Biochem Biophys Res Commun* 170, 963-7.

Morandi, L., Moro, M., Gussoni, E., Tedeschi, S., Correlio, F. 1990. Dystrophin analysis in Duchenne and Becker muscle dystrophy carriers: Correlation with intracellular calcium and albumin. *Ann Neurol* 28, 674-9.

Ohlendieck, K., Matsumura, K., Ionasescu, V.V., *et al.* 1993 Duchenne muscular dystrophy: Deficiency of dystrophin-associated proteins in the sarcolemma. *Neurology* 43, 795-800.

Oudet, C., Heilig, R., Hanauer, A., Mandel, J-L. 1991. Nonradioactive assay for new microsatellite polymorphisms at the 5' end of the dystrophin gene, and estimation of intragenic recombination. *Am J Hum Genet* 49, 311-9.

Passos-Bueno, M.R., Bakker, E., Kneppers, A.L.J., *et al.* 1992. Different mosaicism frequencies for proximal and distal Duchenne muscular dystrophy (DMD) mutations indicate difference in etiology and recurrence risk. *Am J Hum Genet* 51, 1150-5.

Petrof, B.J., Shrager, J.B., Stedman, H.H, Kelly A.M., Sweeney, H.L. 1993. Dystrophin protects the sarcolemma from stresses developed during muscle contraction. *Proc Natl Acad Sci USA* 90, 3710-4.

Powell, J.F., Fodor, F.H., Cockburn, D.J., Monaco, A.P., Craig, I.W. 1991. A dinucleotide repeat polymorphism at the DMD locus. *Nucleic Acids Res* 19, 1159.

Prior, T.W., Friedman, K.J., Silverman, L.M. 1989. Detection of Duchenne/Becker muscular dystrophy carrier by densitometric scanning. *Clin Chem* 35, 1256-7.

Prior, T.W., Friedman, K.J., Highsmith, W.E., Jr., Perry, T.R., Silverman, L.M. 1990a. Molecular probe protocols for determining carrier status in Duchenne and Becker muscular dystrophies. *Clin Chem* 36, 441-5.

Prior, T.W., Papp, A.C., Snyder, P.J., *et al.* 1990b. Determination of carrier status in Duchene and Becker muscular dystrophies by quantitative polymerase chain reaction and allele-specific oligonucleotides. *Clin Chem* 36, 2113-7.

Prior, T.W., Papp, A.C., Snyder, P.J., Mendell, J.R. 1992. Case of the month: Germline mosaicism in carriers of Duchenne muscular dystrophy. *Muscle and Nerve* 15, 960-3.

Prior, T.W., Papp, A.C., Snyder, P.J., *et al.* 1993a. A missense mutation in the dystrophin gene in a Duchenne muscular dystrophy patient. *Nature Genet* 4, 357-60.

Prior, T.W., Papp, A.C., Snyder, P.J., *et al.* 1993b. Identification of two point mutations and a one base deletion in exon 19 of the dystrophin gene by heteroduplex formation. *Hum Molecular Genet* 2, 311-3.

Roberts, R.G., Cole, C.G., Hart, K.A., Bobrow, M., Bentley, D.R. 1989. Rapid carrier and prenatal diagnosis of Duchenne and Becker muscular dystrophy. *Nucleic Acids Res* 17, 811.

Roberts, R.G., Bobrow, M., Bentley, D.R. 1991. A MaeIII polymorphism near the dystrophin gene promoter by restriction of amplified DNA. *Nucleic Acids Res* 18, 1315.

Roberts, R.G., Bobrow, M., Bentley, D.R. 1992. Point mutations in the dystrophin gene. *Proc Natl Acad Sci USA* 89, 2331-5.

Saiki, R.K., Scharf, S., Faloona, F., *et al.* 1985. Enzymatic amplificaton of β-globin genomic sequences and restriction site analysis for diagnosis of sickle cell anemia. *Science* 130, 1350-4.

Schwartz, L.S., Tarleton, J., Popovich, B., Seltzer, W.K., Hoffman, E.P. 1992. Fluorescent multiplex linkage analysis and carrier detection for Duchenne/Becker muscular dystrophy. *Am J Hum Genet* 51, 721-9.

Silverman, L.M., Mendell, J.R., Sahenk, Z., Fontana, M.B. 1976. Significance of creatine kinase phosphokinase isoenzymes in Duchenne dystrophy. *Neurology* 26, 561-4.

Thompson, M.W., McInnes, R.R., Willard, H.F. 1991. Thompson and Thompson. *Genetics in Medicine*. 5th edn. WB Saunders Co, Philadelphia.

Vainzof, M., Pavanello, R.C.M., Pavanello-Filho, I., *et al.* 1991a. Screening of male patients with autosomal recessive Duchenne dystrophy through dystrophin and DNA studies. *Am J Med Genet* 39, 38-41.

Vainzof, M., Pavanello, R.C.M., Pavanello, I., *et al.* 1991. Dystrophin immunofluorescence pattern in manifesting and asymptomatic carriers of Duchenne's and Becker muscular dystrophies of different ages. *Neuromusc Dis* 1, 177-83.

Wilton, S.D., Johnsen, R.D., Pedretti, J.R., Laing, N.G. 1993. Two distinct mutations in a single dystrophin gene: Identification of an altered splice-site as the primary Becker muscular dystrophy mutation. *Am J Med Genet* 46, 563-9.

Winnard, A.V., Klein, C.J., Coovert, D.D., *et al.* 1993. Characterization of translational frame exception patients in Duchenne/Becker muscular dystrophy. *Hum Molecular Genet* 2, 737-44.

Yau, S.C., Roberts, R.G., Bentley, D.R., Mathew, C.G., Bobrow, M. 1991. A Mse1 polymorphism in exon 48 of the dystrophin gene. *Nucleic Acids Res* 196, 5803.

Yau, S.C., Roberts, R.G., Bobrow, M., Mathew, C.G. 1993. Direct diagnosis of carriers of point mutations in Duchenne muscular dystrophy. *Lancet* 341, 273-5.

Zneimer, S.M., Schneider, N.R., Richards, C.S. 1993. In situ hybridization shows direct evidence of skewed X inactivation in one of monozygotic twin females manifesting Duchenne muscular dystrophy. *Am J Med Genet* 45, 601-5.

CHAPTER 8

Trinucleotide Repeat Instability: Clinical Perspectives Using the Fragile X Syndrome and Myotonic Dystrophy as Examples

Thomas W. Prior, Ph.D.

Introduction

The fragile X syndrome and myotonic dystrophy are two genetic disorders that have recently been shown to be due to a new class of gene mutations, often referred to as dynamic mutations. The mutation mechanism involves normally occurring polymorphic trinucleotide repeats in the respective genes which, when expanded beyond the normal range, result in a loss of function of the gene. The rate of mutation is related to the copy number of the repeats; thus, as the repeat expands in size over successive generations, the mutation rate actually changes. All previously described mutations in human genes, such as deletions, point mutations, or duplications, have been static, in that the mutant sequence has exhibited the same rate of mutation as its predecessor (Richards and Sutherland 1992a). Furthermore, the atypical segregation patterns observed both in individuals with the fragile X syndrome and with myotonic dystrophy (referred to as the Sherman paradox and anticipation, respectively) can now be explained in terms of changes in the inherited unstable repeat size. Both diseases exhibit variable expressivity, but this is not the result of different mutations in the genes; with few exceptions, the mutations in all fragile X and myotonic dystrophy patients involve the same trinucleotide repeats in each gene. Mechanistic considerations and the clinical implications of the recent gene findings for the fragile X syndrome and for myotonic dystrophy will be discussed.

The Fragile X Syndrome

Description of the Fragile X Syndrome

The fragile X syndrome is the most common cause of familial mental retardation, having an incidence of approximately one in 1500 males and one in 2000–2500 females (Turner et al. 1986). This unique syndrome gets its unusual name from a fragile site on the tip of the long arm of the X chromosome (band q27.3) in metaphase cells from affected patients (Lubbs 1969). The fragile sites appear as an unstained gap or break on the X chromosome by cytogenetic analysis. This fragile site is not expressed constitutively, but must be induced by culturing cells from whole blood, amniotic fluid or chorionic villi under conditions of folate deprivation or imbalance of deoxynucleotide synthesis. However, the cytogenetic test has limitations. The analysis is tedious and the site is usually only observed in 10–40% of the metaphases in affected males and less frequently in retarded females (Richards and Sutherland 1992b). Interpretation of the cytogenetic results has also been ambiguous in cases when a low percentage of sites is observed. The cytogenetic marker is a relatively good test for the detection of most retarded males and females; however the fragile site is detectable in only about 55% of obligate carriers (Sherman et al. 1984), thus making it an unreliable test for the accurate determination of carrier status.

The clinical symptoms in males include moderate mental retardation, coarse facial features with long jawbones, a high wide forehead, large ears, macro-orchidism, mild connective tissue abnormality that results in fine skin, and hyperextensible joints (Nussbaum and Ledbetter 1989). Affected females are often less severely mentally retarded and may show some craniofacial and connective tissues features. The physical symptoms are often more apparent after early childhood. The stereotypical fragile X behavior is somewhat hyperactive, especially in younger children, with poor social interactions. Some of the affected children also display autistic-like behavior. However, this syndrome exhibits genetic variability, with some retarded individuals exhibiting few of the other symptoms. Clinical diagnosis of the fragile X syndrome is often not straightforward, particularly in young children.

The inheritance of fragile X is seemingly paradoxical. It is considered to be an X-linked dominant disorder with incomplete penetrance; both sexes can exhibit mental retardation, and approximately 20% of males who must have the fragile X chromosome because they have affected grand-children are phenotypically normal and do not express a fragile site at Xq27 (Sherman et al. 1985). These unaffected males, referred to as normal transmitting males (NTMs), transmit the mutation to all of their daugh-ters, all of whom are also asymptomatic. However, the members of the next generation, the NTMs' grandchildren, are often mentally impaired. Sherman et al. determined that the risk of mental retardation is increased by the number of generations through which the mutation has passed; this is known as the Sherman Paradox (Sherman et al. 1985) . The grand-sons of NTMs have a risk of 40% of mental deficiency, and great-grandsons are at 50% risk. Conversely, 30% of known carrier females have some degree of mental impairment, and the risk that their sons will be affected is 50%. These findings have made genetic counseling for fragile X very difficult. However, the recent identification of the fragile X mutation and the advent of direct DNA testing have enhanced our understanding of this unusual inheritance and has improved our ability to predict the phe-notype.

The Fragile X Gene Mutation

Recent studies have shown that the fragile X mutation is the result of an unstable DNA sequence in a gene called FMR-1 for fragile X mental retardation-1 (Verkerk et al. 1991; Yu et al. 1991; Oberle et al. 1991). The disease pathogenesis is a multistep process involving two gene changes: length mutations of a CGG repeat and aberrant methylation of the CpG island adjacent to FMR-1. At the 5' end of the gene there is a (CGG)n repeat sequence, which is located 250 bp downstream from the CpG island. The repeat is polymorphic and ranges in size from 6 to 54 triplet repeats, with a mean of 29, on normal X chromosomes (Fu et al. 1991). It is unclear whether the CGG sequence is translated into protein (Nelson 1993); however, the repeat is the site of the mutation and is directly proportional to the increase in size of the unstable region of DNA (Kremer et al. 1991). In normal transmitting males and all of their daughters, small expansions of approximately 60–230 repeats are observed. These amplifications are referred to as premutations and are not associated with clinical symptoms but, in contrast to normal alleles, often change in size during meiotic division. Furthermore, the premutations always precede the appearance of full mutations which are associated with the fragile X phenotype. The direct passage from a normal allele to a full mutation has not been reported. Full mutations cause mental retardation in all males, in about 50% of females and are found in some carrier females. The full mutation not only involves further amplification (> 230 repeats), it also involves methylation of the 5' CpG island. CpG islands are often found near transcriptional start sites, and specific methylation of the CpG dinucleotide sequence has been shown to turn off the expression of adjacent genes (Toniolo et al. 1988). Therefore, the consequence of the transition from the premutation to full mutation appears to be loss of the expression of FMR-1 (Pieretti et al. 1991). Premutation to full mutation changes are found only in the offspring of females, and the risk of expansion to full mutations is directly proportional to the size of the premutation in the mother. Male transmission of the gene is accompanied by only small changes of the repeat within the premutation range. These properties combine to account for the Sherman paradox; the general increase in size of the premutation with successive generations when transmitted through females increases the risk of mothers having mentally handicapped children.

Diagnostic Testing

The changes in both the CGG repeat length and the methylation status of the CpG island can be readily monitored by Southern blots after DNA digestion with a combination of two restriction enzymes (Oberle et al. 1991; Rousseau et al. 1991; Yu et al. 1991). At The Ohio State University, our current protocol uses probe StB12.3 (Oberle et al. 1991) which is adjacent to the repeat, and HindIII + EagI digests of patient DNA. The probe detects a 2.8 kb HindIII-EagI fragment corresponding to the normal active X in males and females and an additional 5.2 kb fragment corresponding to the normal inactive X in females. EagI does not cut the CpG site when it is methylated on the inactive X chromosome; therefore digestion results in an uncut 5.2 kb fragment.

In affected males with full mutations, the size of the DNA fragment containing the CGG repeat is markedly increased and the CpG island is

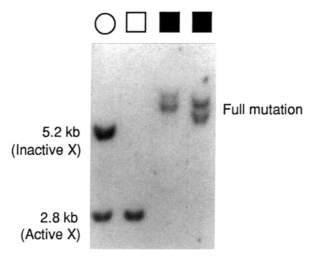

Figure 8.1. Detection of fragile X full mutations. Squares and circles represent males and females. Solid symbols represent fragile X mutation-positive mentally retarded individuals. Leukocyte DNA was simultaneously digested to completion with HindIII and EagI and hybridized with probe StB12.3.

methylated. This corresponds to fragments larger than 5.2 kb. An increase in the range of 1–3 kb is generally observed. In Figure 8.1, as shown in both of the affected patients, a heterogeneous pattern of methylated and expanded fragments is often associated with full mutations. The diffuse smear of DNA fragments, which is quite common with the full mutation pattern, indicates somatic mosaicism—different cells having fragments with different numbers of the CGG repeat. Occasionally, affected males will carry not only large methylated DNA sequences, but also smaller non-methylated DNA sequences as well (Rousseau *et al.* 1991). We have observed this mosaic pattern in about 10% of our patient population. It has been suggested that mosaic males may be more mildly retarded than males with only full mutations. However, it is important to realize that correlations between the degree of mental retardation and presence of unmethly-ated repeats are being established from studies where lymphocyte DNA is analyzed. Since brain cells are most likely to be involved in the expression of the syndrome and somatic mosaicism is a common feature, geno-type/phenotype correlations may be more complicated and less accurate than in other genetic disorders.

We have studied several families where the initial carrier of the pre-mutation was a male. Premutations in males are unmethylated and are

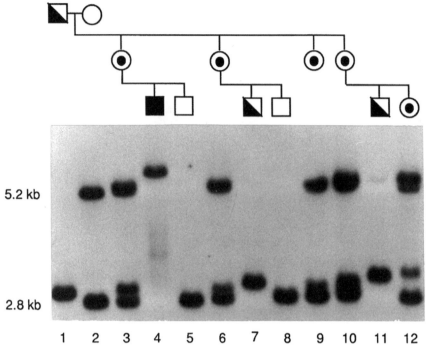

Figure 8.2. Analysis of a fragile X affected family. Squares and circles represent males and females. Half-filled squares represent normal transmitting males. Dotted circles represent carrier females. Leukocyte DNA was simultaneously digested to completion with HindIII and EagI and hybridized with probe StB12.3.

seen as an expansion of the 2.8 kb fragment by 70 to about 500 bp. The number of repeats stays in the premutation range and shows minimal alteration when transmitted from males to their daughters. An example of a family containing three normal transmitting males is shown in Figure 8.2. As shown, the repeat sequence is relatively stable in size when transmitted from the father (lane 1) to each of his four carrier daughters (lanes 3, 6, 9 and 10). No large expansions from male premutations have been observed; hence daughters of NTMs are all obligate carriers but are not at risk for having the syndrome. In the carrier females, the presence of premutations yields a typical 4 band pattern which represents the four possible combinations of normal active (2.8 kb), premutated active (2.8 kb + Δ), normal inactive (5.2 kb), and premutated inactive (5.2 kb + Δ) genes. These bands often have variable intensities. However, the premutation fragments are better resolved above the 2.8 kb fragment than the 5.2 kb fragment, and therefore a 3 band pattern is more commonly observed. When transmitted by the carrier females, the repeat sequence usually increases and may vary in the exact size of the expansion. In Figure 8.2, one of the carrier daughters (lane 3) had an affected son with a full mutation (lane 4), while two of the carrier daughters (lanes 6 and 9) had NTM offspring (lanes 7 and 11). As shown in this family, the amount by which the repeat expands and when methylation takes place is difficult to predict.

The phenomenon of a fragment increasing in size from one generation to the next is often observed in fragile X families. Although the precise mechanisms of amplification and methylation remain unclear, it has been shown that the risk of expansion to full mutation correlates with the size of the premutation allele. The risk is low in the 60 repeat range, is intermediary in the 70 repeat range, and reaches 100% when above 100 repeats (Fu et al. 1991). This relationship further accounts for the Sherman paradox. Also, this has important consequences in genetic counseling, since the risk of having affected children will vary considerably depending upon the size of the premutation in the female carrier. Although repeat sizes in the normal and premutation DNA fragments can be accurately determined by PCR (Fu et al. 1991; Pergolizzi et al. 1992), distinguishing between large normal alleles and small premutations can be problematic. According to Fu et al., CGG repeat sizes of 46 or less are meiotically stable, while repeat sizes of 52 and greater are unstable (Fu et al. 1991). The CGG repeat can be interrupted by one or two AGG repeats in normal individuals (Verkerk et al. 1991). It has been proposed that the presence of AGG repeats may be important in maintaining the stability of the CGG repeat. The relationship between repeat size and stability must be firmly established before one can consider initiating population carrier screening for this disorder.

Figure 8.3 shows the pattern of a mildly retarded fragile X positive female (lane 3). A large methylated band coexists with the normal methylated and unmethylated bands. Approximately 50% of females with full mutations are mentally impaired (Rousseau et al. 1991). As a consequence, it is not possible to predict the phenotype of females with full mutations from DNA studies. In the affected females, it is speculated that that FMR-1 expression is impaired by preferential inactivation of the normal allele in brain tissue.

Due to the cloning of the fragile site, fragile X diagnostic testing has significantly improved. The new direct DNA testing is particularly useful for the identification of affected males and the determination of carrier status in affected families. The DNA studies have dramatically helped clinicians by providing an accurate and relatively inexpensive method for the differential diagnosis of mental retardation. The test can be used to screen high risk individuals with mental retardation, learning and developmental disabilities, and autism. To date, no new mutations have been reported in fragile X (Yu *et al.* 1992; Smits *et al.* 1993). All isolated affected individuals have been shown to have a carrier parent. The lack of new mutations implies that it may take a number of generations before the disease manifests itself in a family and, as a result, there may be more carriers than previously anticipated. Since all cases of the disease have been shown to be familial, accurate carrier studies are extremely important and are indicated throughout extended affected families (Smits *et al.* 1993). Since the cytogenetic analysis for the determination of carrier status has a high false negative rate (approximately 45%), and false positives may also occur when a low frequency of fragile sites are expressed, DNA testing has now become the standard method for identi-

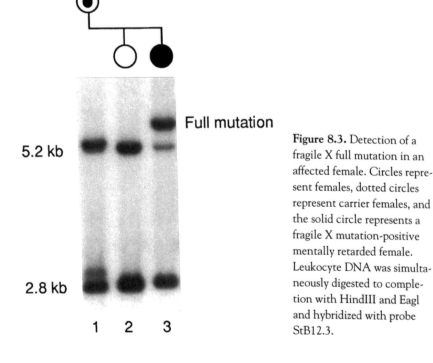

Full mutation

5.2 kb

2.8 kb

1   2   3

Figure 8.3. Detection of a fragile X full mutation in an affected female. Circles represent females, dotted circles represent carrier females, and the solid circle represents a fragile X mutation-positive mentally retarded female. Leukocyte DNA was simultaneously digested to completion with HindIII and EagI and hybridized with probe StB12.3.

fying carriers of the fragile X syndrome. At our institution, we are no longer performing fragility testing. However, cytogenetic karotype analysis still has a very important function in evaluating isolated cases of developmental delay, since other chromosomal anomalies may be detected. Our current protocol includes a chromosome study on all fragile X negative samples. We have found that approximately 5% of the fragile X mutation negative patients evaluated for developmental delay and other forms of mental retardation have other cytogenetic anomalies.

Myotonic Dystrophy

Description of Myotonic Dystrophy

Myotonic dystrophy (DM) is the most common inherited form of muscular dystrophy affecting adults, having an incidence of approximately 1 in 8000 individuals (Harper 1989). The disease is characterized by progressive muscle weakness and myotonia, resulting in delayed muscle relaxation. However, unlike many of the other muscular dystrophies, DM is often a multisystem disorder in which other symptoms commonly accompany the disorder, including cataracts, conduction defects of the heart, intellectual impairment, a form of diabetes, and in some males, testicular atrophy and frontal balding (Harper 1989). The median age of onset is about 20–25 years of age, although extreme variation has been observed between and within families. Electromyography to confirm myotonia, slit lamp examination for retinal changes and cataracts, and a moderately elevated creatine kinase are tests that can assist in the diagnosis of DM (Harper 1989b). However, the diagnosis can be problematic because of the wide range and severity of symptoms, and often affected individuals will already have had children before they have been diagnosed. There is also a severe congenital form of the disorder which results in mental retardation, respiratory distress, hypotonia, and in many cases, death shortly after birth due to respiratory complications. The congenital form is seen only in the offspring of women who are themselves mildly affected (Harper 1975).

Myotonic dystrophy is inherited as an autosomal dominant trait and is characterized by a highly variable expression, ranging from neonatal mortality in congenital cases to extremely mild forms which are seen in middle or old age and are characterized by cataracts and little or no muscle involvement. Although variation in expression is severe, the gene appears to be fully penetrant. The genetic phenomenon of anticipation has been reported in DM families (Harper 1989; Howeler et al. 1989). Anticipation denotes progressively earlier appearance of a disease in successive generations, generally with increasing severity. The phenomenon is most apparent in those families with congenitally affected children. Penrose (1948) attributed this phenomenon to ascertainment bias, a bias in the observation and recording of multigeneration families with mild features in older generations and earlier onset in successive generations. However, the recent identification of the DM mutation provides strong evidence for the existence of genetic anticipation in the transmission of DM.

The Myotonic Dystrophy Gene Mutation

Using genetic linkage analysis and molecular techniques, the gene for DM was initially mapped to the proximal long arm of chromosome 19 (Whitehead et al. 1982; O'Brien et al. 1983). The DM mutation was recently identified, and, as found in the fragile X syndrome, the basis of the mutation was an unstable triplet repeat (Aslanidis et al. 1992; Brook et al. 1992; Buxton et al. 1992; Fu et al. 1992; Mahadevan et al. 1992). A polymorphic CTG repeat is present in the normal population and ranges in size from 5 to 35 repeats (Davies et al. 1992). In DM-affected patients, the number expands from 50 to several thousand. Similarly to the fragile X syndrome, the copy number of the DM mutant allele tends to increase in successive generations, and the expanded repeats are both meiotically and mitotically unstable. However, in DM there appears to be a higher degree of positive correlation between repeat length and clinical severity than observed in fragile X families. The congenital cases have been shown to have the largest expansions, between 730 and 4300 copies of the repeat, whereas the mild and subclinical cases (no symptoms other than cataracts or nondiagnostic electromyogram) have been shown to have smaller repeat sizes in the 50–100 range (Redman et al. 1993). Although there was overlap between the repeat size and the age of onset, an overall correlation was recently reported (Hunter et al. 1992). This trend of greater clinical severity with increasing expansion size was also apparent when school performance was examined (Hunter et al. 1992). However, due to the somatic heterogeneity seen in DM, phenotype/genotype correlations derived from peripheral blood may not be as accurate as the measurement of the repeat size in the actual affected tissue.

Several studies now support the existence of anticipation in DM, which can be explained in terms of progressive DNA expansions over successive generations. In the congenital cases, it has been shown that mothers with larger repeats have a higher risk of transmitting an expanded congenital allele than mothers with smaller allele size (Tsilifidis *et al.* 1992; Redman *et al.* 1993). Although the size of the expansion is influenced by the sex of the transmitting parent, with the congenital cases only occurring through maternal transmission, high risk alleles can expand via both male and female meioses. This differs from fragile X, where repeat amplifications are limited to female meioses.

It has further been found that the CTG repeat is located within the 3' untranslated region of a gene that appears to encode a protein kinase, named myotonin protein kinase (Brook *et al.* 1992; Fu *et al.* 1992, Mahadevan *et al.* 1992). Since protein kinases are involved in signal transduction pathways in all cells in the body, a defective protein kinase may explain how a single gene defect could result in the diverse effects observed in the disease. The molecular mechanism by which the mutation exerts its dominant expression of the DM phenotype is unclear. Unlike fragile X, there have been no findings of abnormal methylation events associated with the DM expansion. A recent study, using quantitative reverse transcription-polymerase chain reaction and radioimmunoassay, demonstrated that decreased levels of messenger RNA and protein expression were associated with the adult form of myotonic dystrophy (Fu *et al.* 1993). It was suggested that the dominant nature of DM was likely to be the result of a dosage-dependant mechanism, whereby a reduction of myotonin kinase causes a disruption in signal transduction pathway. Earlier biochemical studies had shown reduced phosphorylation of membrane proteins from DM patients (Roses and Appel 1973). However, in another report, expression studies on samples from congenital cases demonstrated marked increases in the messenger RNA steady state levels (Sabouri *et al.* 1993). In contrast to a reduction of myotonin kinase, the authors proposed that the effect of the mutation would be a non-regulated hyperphosphorylation of its substrate by high levels of myotonin kinase. Further studies are necessary to define precisely the mechanism by which GCT amplifications lead to DM in adults or newborns.

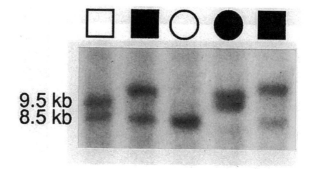

Figure 8.4. Detection of myotonic dystrophy mutations. Squares and circles represent males and females. Solid symbols represent myotonic dystrophy affected individuals. Leukocyte DNA was digested to completion with HindIII and hybridized with probe pMYDI.

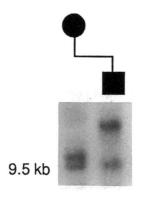

Figure 8.5. Detection of a congenitally affected myotonic dystrophy male. Squares and circles represent males and females. Solid symbols represent myotonic dystrophy-affected individuals. The affected son was an infant with severe congenital myotonic dystrophy. Leukocyte DNA was digested to completion with HindIII and hybridized with probe pMDYI.

Diagnostic Testing

The GTC expansion can be readily monitored by Southern blots after DNA digestion with a number of different restriction enzymes, including BamHI, EcoRI, BglI, PstI, SacI, and HindIII. Enzymes generating smaller restriction fragments containing the repeat sequence increase the resolution and the ability to identify minimal expansions. The majority of clinically significant mutations can probably be identified by Southern analysis; however, for small amplifications (< 100 repeats), PCR analysis is essential. Our current Southern protocol uses the probe pMDY1 (Fu et al. 1992), which spans the repeat area, and HindIII restriction digestion. Probe pMDYI detects a HindIII polymorphism with alleles of 9.5 and 8.5 kb alleles (or 10.0 and 9.0 kb EcoRI alleles), the frequencies of which are approximately 0.60 and 0.40 respectively. The larger allele is due to a 1 kb insertion telomeric to the CTG repeat, and it has been shown that the larger allele is in almost complete linkage disequilibrium with the mutation (Mahadevan et al. 1992). This suggests that there were a limited number of mutations which occurred on a chromosome having the larger allele. Alternatively, the larger allele may be predisposed to DM mutations.

The genomic fragments detected by pMDYI from normal controls and unrelated individuals affected with DM are shown in Figure 8.4. Typical increases in the range of 1—4 kb are observed on Southern blots containing DNA from individuals affected with DM. Many of the larger expansions are detected as smears, indicating somatic cell heterogeneity of the expanded alleles similar to that seen in fragile X. Figure 8.5 shows a DM congenital case, showing one of the larger expansions (> 4 kb) we have observed on our patient population. The congenital cases are exclusively maternally transmitted, and the likelihood that a mother will have an infant with the congenital form correlates with the expansion size. Redman et al. (1993) found that all of the severe congenital cases had repeat sizes greater than 730 repeats, and mothers with 100 repeats or more have a 62% risk of transmitting a congenital DM allele, while those with fewer than 100 repeats have a 36% risk (Redman et al. 1993). Since CTG expansion is not gender dependent, but the congenital expansions are, it implies that other factors in maternal transmission must be important, such as imprinting, maternal genes or intrauterine factors.

Figure 8.6 shows the genetic instability of the DM locus. The father showed no muscle weakness but had undergone cataract surgery. Progressive enlargements of the HindIII fragments occurred during the transmission of the mutation to each of his affected sons. The sons are presently in their late twenties, and both exhibit facial weakness with bilateral ptosis

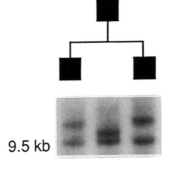

9.5 kb

Figure 8.6. Analysis of a myotonic dystrophy-affected family. Solid squares represent myotonic dystrophy-affected males. Leukocyte DNA was digested to completion with HindIII and hybridized with probe pMDYI.

and tongue myotonia. The molecular analysis confirmed the clinical diagnosis in both sons. The DNA analysis is also helpful in determining the origin of the mutation when there is a negative family history and the parents are asymptomatic or exhibit equivocal symptoms, such as cataracts. No new mutations have been described in DM, which is consistent with the linkage disequilibrium data. In order to account for the maintenance of the mutation in the population, it was recently proposed that there is a high incidence of minimally expanded alleles in DM families which produce few symptoms and are stably transmitted over several generations (Barcelo et al. 1993). Therefore, for counseling purposes, it becomes important to identify in which side of the family the mutation is segregating. When comparing unrelated affected individuals with small to moderate differences in repeat sizes, it is generally difficult to accurately predict the severity of the disease in each case. However, when a child has a significant increase in allele size when compared to the parent, as is the case with both of the sons in Figure 8.6, it is almost certain that there will be an earlier age of onset and more severe disease.

Several cases of "reverse" mutations have now been reported in DM (Brunner et al. 1993; Hunter et al. 1993; O'Hoy et al. 1993), wherein there is a spontaneous correction of a deleterious mutation upon transmission to an unaffected offspring. Although small reductions in the repeat region have been found in individuals with the fragile X syndrome repeat, there have been no cases of a reversal from an affected to a normal allele with a corresponding change in phenotype. The mechanism for the DM reverse mutations remains unknown, although the possibility of a single recombinational event was ruled out in these reports. A gene conversion mechanism, whereby the normal parental allele replaces the expanded allele, has not been reported in humans but may best account for the occurrence in DM. The reversions may provide an explanation of the nonpenetrance observed in some DM families.

The discovery of an expanded repeat sequence in myotonic dystrophy has greatly improved our ability to detect DM carriers who are asymptomatic or show few of the classical signs of the disease. Although closely linked RFLP markers have been available for several years, linkage analysis is not as accurate as direct DNA testing for the mutation, especially when there is an uncertain diagnosis or when key family members are unable to be tested. Also, linkage results provide no information regarding the sever-

Table 8.1 Features of diseases caused by trinucleotide repeat expansions

	Fragile X syndrome	Mytonic dystrophy	Huntington disease	Spinobulbar muscular atrophy	Spino-cerebellar ataxia type I
Chromo-some locus	Xq27.3	19q13.3	4p16.3	Xq21.3	6p24
Tri-nucleotide repeat unit	CGG in 5' UTR[a] of FMR-1 gene (unknown function)	CTG in 3' UTR[a] of a cAMP-dependent muscle protein kinase	CAG repeat in coding region of huntingin gene (unknown function)	CAG in coding region of androgen receptor gene	CAG in coding region of gene of unknown function
Normal range of repeats	6–54	5–35	10–36	13–30	19–36
Premutation range	50–200	–	–	–	–
Disease range	> 230	44–> 4000	42–100	40–62	43–81
Sex bias in transmission of disease allele	Expansion during maternal meioses only.	Expansion in both males and females. Severe form transmitted by mother.	Paternal transmission of severe juvenile form		

[a] untranslated region

ity of the disease. It is no longer necessary for patients to undergo muscle biopsy, muscle enzyme /tudy, and electromyography as the first diagnostic procedure. We have found that the majority of the mutations in affected patients can be detected by conventional Southern analysis, but for mildly affected patients with small expansions, more sensitive PCR-based assays should be used.

Conclusions

The dynamic type of mutation found in the fragile X syndrome and myotonic dystrophy have also been identified in spinal and bulbar muscular atrophy (La Spada *et al.* 1991), Huntington disease (The Huntington Disease Collaborative Research Group 1993), and spinocerebellar ataxia type 1 (Orr *et al.* 1993). It is anticipated that other genetic diseases which do not follow classical Mendelian rules and exhibit some of the unusual genetic properties described in the fragile X syndrome and myotonic dystrophy (such as the variable expression and unusual segregation patterns) will also be due to heritable unstable DNA sequences. Although there are properties shared by diseases associated with unstable elements, there are also important differences: composition of the repeat, size of the repeat, location of the repeat, and consequence of the repeat on the expression of the gene. Some of the differences are shown in Table 8.1.

When the basic mechanism of what makes the DNA sequences unstable is elucidated, then possible therapeutic strategies will be developed for these diseases. However, the discovery of these mutations is already having a major impact on diagnostic testing. The application of direct DNA-based assays in fragile X and myotonic dystrophy has significantly improved the accuracy of diagnosis and has provided families with more accurate risk estimates. Today, through genetic counseling, at-risk family members are able to make family planning decisions with information that was not available a short time ago.

References

Aslanidis, C., Jansen, G., Amemiya, C., *et al.* 1992. Cloning of the essential myotonic dystrophy region and mapping of the putative defect. *Nature* 355:548-51.

Barcelo, J.M., Mahadevan, M.S., Tsilfidis, C., MacKenzie, A.E., Korneluk, R.G. 1993. Intergenerational stability of the myotonic dystrophy promutation. *Hum Mol Genet* 2:705-09.

Brook, J.D., McCurrach, M.E., Harley, H.G., *et al.* 1992. Molecular basis of myotonic dystrophy: Expansion of a trinucleotide (CTG) repeat at the 3' end of a transcript encoding a protein kinase family. *Cell* 68:799-808.

Brunner, H.G., Jansen, G., Nillesen, W., *et al.* 1993. Brief report: Reverse mutation in myotonic dystrophy. *N Engl J Med* 328:476-80.

Buxton, J., Shelbourne, P., Davies, J., *et al.* 1992. Detection of an unstable fragment of DNA specific to individuals with myotonic dystrophy. *Nature* 355:547-48.

Davies, J., Yamagata, H., Shelbourne, P., *et al*. 1992. Comparison of myotonic dystrophy associated CTG repeat in European and Japanese populations. *J Med Genet* 29:766-69.

Fu, Y.H., Friedman, D.L., Richards, S., *et al*. 1993. Decreased expression of myotonin-protein kinase messenger RNA and protein in adult form of myotonic dystrophy. *Science* 260:235-38.

Fu, Y.H., Kuhl, D.P.A., Pizzuti, A., *et al*. 1991. Variation of the CGG repeat at the fragile X site results in genetic instability: Resolution of the Sherman paradox. *Cell* 67:1047-58.

Fu, Y.H., Pizzuti, A., Fenwick, R.G., *et al*. 1992. An unstable triplet repeat in a gene related to myotonic muscular dystrophy. *Science* 255:1256-58.

Harper, P.S. 1975. Congenital myotonic dystrophy in Britain, II: Genetic basis. *Arch Dis Child* 50:514-521.

Harper, P.S. 1989. *Myotonic Dystrophy*. London, England: WB Saunders.

Harper, P.S. 1989b. The muscular dystrophies. In *Metabolic Basis of Inherited Disease*, Vol I, edited by Scriver, C.R., Beaudet, A.L., Sly, W.S., Valle, D. McGraw-Hill, New York: pp. 2888-2898.

Howeler, C.J., Busch, H.F.M., Geraedts, J.P.M., Niermeij, M.F., Staal, A. 1989. Anticipation in myotonic dystrophy: fact or fiction. *Brain* 112:779-797.

Hunter, A.G.W., Tsilfidis, C., Mettler, G., *et al*. 1992. The correlation of age of onset with CTG trinucleotide repeat amplification in myotonic dystrophy. *J Med Genet* 29:774-79.

Hunter, A.G.W., Jacob, P., O'Hoy, K., *et al*. 1993. Decrease in the size of the myotonic dystrophy CTG repeat during transmisssion from parent to child: Implications for genetic counselling and genetic anticipation. *Am J Med Genet* 45:401-07.

Kremer, E.J., Pritchard, M., Lynch, M., *et al*. 1991. Mapping of DNA instability at the fragile X to a trinucleotide repeat sequence p(CGG)n. *Science* 252:1711-14.

La Spada, A.R., Wilson, E.M., Lubahn, D.B., Harding, A.E., Fishbeck,

K.H. 1991. Androgen receptor gene mutations in X-linked spinal and bulbar muscular atrophy. *Nature* 352:77-79.

Lubbs, H.A. 1969. A marker X chromosome. *Am J Hum Genet* 21:231-44.

Mahadevan, M., Tsilfidis, C., Sabourin, L., *et al.* 1992. Myotonic dystrophy mutation: An unstable CTG repeat in the 3' untranslated region of the gene. *Science* 255:1253-55.

Nelson, D.L. 1993. Fragile X syndrome: Review and current status. *Growth, Genetics, and Hormones* 9:1-4.

Nussbaum, R.L., Ledbetter, D.H. 1989. The fragile X syndrome. In *Metabolic Basis of Inherited Disease*, Vol I, edited by Scriver, C.R., Beaudet, A.L., Sly, W.S., Valle, D. McGraw-Hill, New York: pp. 327-41.

Oberle, I., Rousseau, F., Heitz, D., *et al.* 1991. Instability of a 550 base pair DNA segment and abnormal methylation in fragile X syndrome. *Science* 252:1097-1102.

O'Brien, T., Ball, S., Sarafarazi, M., Harper, P.S., Robson, E.B. 1983. Genetic linkage between the loci for myotonic dystrophy and pepidase P *Ann Hum Genet* 47:117-21.

O'Hoy, K.L., Tsilifidis, C., Mahadevan, M.S., *et al.* 1993. Reduction in size of the myotonic dystrophy trinucleotide repeat mutation during transmission. *Science* 259:809-12.

Orr, H.T., Chung, M.Y., Banfi, S., *et al.* 1993. Expansion of an unstable trinucleotide CAG repeat in spinocerebellar ataxia type 1. *Nature Genet* 4:221-26.

Penrose, L.S. 1948. The problem of anticipation in pedegrees of dystrophia myotonica. *Ann Eugen* 14:125-232.

Pergolizzi, R.G., Erster, S.H., Goonewardena, P., Brown, W.T. 1992. Detection of full fragile X mutation. *Lancet* 339:271-272.

Pieretti, M., Zhang, F., Fu, Y-H., *et al.* 1991. Absence of expression of the FMR-1 gene in fragile X. *Cell* 66:817-22.

Redman, J.B., Fenwick, R.G., Fu, Y.H., Pizzuti, A., Caskey, C.T. 1993. Relationship between parental trinucleotide GCT repeat length and severity of myotonic dystrophy in offspring. *JAMA* 269:1960-65.

Richards, R.I., Sutherland, G.R. 1992a. Dynamic mutations: a new class of mutations causing human disease. *Cell* 70: 709-12.

Richards, R.I., Sutherland, G.R. 1992b. Fragile X syndrome: the molecular picture comes into focus. *Trends Genet* 8:249-253.

Roses, A.D., Appel, S.H. 1973. Protein kinase activity in erythrocyte ghosts of patients with myotonic muscular dystrophy. *Proc Natl Acad Sci USA* 70:1855-59.

Rousseau, F., Heitz, D., Biancalana, V., *et al.* 1991. Direct diagnosis by DNA analysis of the fragile X syndrome of mental retardation. *N Engl J Med* 325:1673-1681.

Sabouri, L.A., Mahadevan, M.S., Narang, M., Lee, D.S.C., Surh, L.C., Korneluk, R.G. 1993. Effect of the myotonic dystrophy (DM) mutation on mRNA levels of the DM gene. *Nature Genet* 4:233-38.

Sherman, S.L., Morton N.E., Jacobs P.A., Turner, G. 1984. The marker X chromosome: A cytogenetic and genetic analysis. *Am J Hum Genet* 48:21-37.

Sherman, S.L., Jacobs P.A., Morton N.E., *et al.* 1985. Further segregation of the fragile X syndrome with special reference to transmitting males. *Hum Genet* 69:3289-99.

Smits, A.P.T., Dreesen, J.C.F., Post, J.G., *et al.* 1993. The fragile X syndrome: No evidence for any new mutations. *J Med Genet* 30:94-96.

The Huntington's Disease Collaboration Research Group. 1993. A novel gene containing a trinucleotide repeat that is expanded and unstable on Huntington's Disease Chromosomes. *Cell* 72:971-83.

Toniolo, D., Martini, G., Migeon, B.R., Dono, R. 1988. Expression of the G6PD locus on the human X chromosome is associated with demethylation of three CpG islnds within 100 kb of DNA. *EMBO J* 7:401-6.

Tsilifidis, C., MacKenzie, A.E., Mettler, G., Barcelo, J., Korneluk, R.G. 1992. Correlation between CTG trinucleotide repeat length and frequency of severe congenital myotonic dystrophy. *Nature Genet* 1:192-95.

Turner, G., Robinson, H., Laing, S., Purvis-Smith, S. 1986. Preventive screening for the fragile X syndrome. *N Engl J Med* 315:607-9.

Verkerk, A.J.H.M., Pieretti, M., Sutcliffe, J.S., *et al.* 1991. Identification of a gene (FMR-1) containing a CGG repeat coincident with a fragile X breakpoint cluster region exhibiting length variation in fragile X syndrome. *Cell* 65:905-14.

Whitehead, A.S., Solomon, E., Chambers, S., Bodmer, W.F., Povey, S., Fey, G. 1982. Assignment of the structural gene for the third component of human complement to chromosome 19. *Proc Natl Acad Sci USA* 79:5021-25.

Yu, S., Pritchard, M., Kremer, E., *et al.* 1991. Fragile X phenotype characterized by an unstable region of DNA. *Science* 252:1179-81.

Yu, S., Mulley, J., Loesch, D., *et al.* 1992. Fragile-X syndrome: unique genetics of the heritable unstable element. *Am J Hum Genet* 50:968-80.

CHAPTER 9

DNA Analysis of Inherited Genetic Changes: Counseling and Social Issues

Nancy Callanan, M.S.

Introduction

DNA-based genetic testing is performed under a variety of circumstances: diagnostic testing; carrier testing; prenatal diagnosis; presymptomatic testing; and susceptibility testing. Important components of any genetic testing are clinical genetics evaluation and genetic counseling. This chapter will discuss the various types of genetic testing and the genetic counseling issues raised by each type of testing. Ethical and social considerations of genetic testing will also be explored.

Genetic Counseling

What Is Genetic Counseling?

In 1975 a committee of the American Society of Human Genetics developed the following definition of genetic counseling:

> Genetic counseling is a communication process that deals with the human problems associated with the occurrence, or the risk of occurrence, of a genetic disorder in a family. This process involves an attempt by one or more appropriately trained persons to help the individual or family (1) comprehend the medical facts, including the diagnosis, the probable course of the disorder, and the available management; (2) appreciate the way heredity contributes to the disorder, and the risk of recurrence in specified relatives; (3) understand the options for dealing with the risk of recurrence; (4) choose the course of action

which seems appropriate to them in view of their risk and their family goals and act in accordance with that decision; and (5) make the best possible adjustment to the disorder in an affected family member and/or to the risk of recurrence of that disorder. (ad hoc Committee on Genetic Counseling 1975)

Geneticists and genetic counselors today still adhere to this basic definition and to the tradition of providing genetic counseling in a non-directive manner. Generally, genetic counselors view their role as providing education and information, exploring the available options and facilitating decision making. In 1991 the National Society of Genetic Counselors adopted a Code of Ethics that further defines the relationship of the genetic counselor and client as one in which the counselor strives to protect the autonomy of the client and to protect the rights of a client to make informed decisions free of coercion (Journal of Genetic Counseling 1992).

Who Provides Genetic Counseling?

In the United States clinical genetic services are generally provided by a team of professionals that includes physicians trained in Medical Genetics, doctoral level clinical geneticists, masters level genetic counselors, nurses, and social workers. The American Board of Medical Genetics (ABMG) was established in 1982 as the national certification board for genetics professionals. The ABMG currently grants certification to qualified M.D. Clinical Geneticists, Ph.D. Medical Geneticists, Biochemical Geneticists, Cytogeneticists, and Molecular Geneticists. Until recently the ABMG also provided certification for non-doctoral level genetic counselors. As of 1993 these professionals will be certified by the American Board of Genetic Counseling.

Why Do People Seek Genetic Counseling?

Individuals and families are referred for or request on their own genetics evaluation and counseling for a variety of reasons. The goal of a clinical genetics evaluation is to first establish or confirm the genetic diagnosis. Once this is accomplished, appropriate information and genetic counseling can be provided to the family. The importance of an accurate diagnosis cannot be overemphasized. The task of establishing an accurate diagnosis is complicated by the nature of many genetic conditions. While it is beyond the scope of this chapter to provide a full discussion of these issues, it is important to note that in the process of establishing a diagnosis, clinical geneticists are ever mindful of the variable expression of many genetic disorders and of the fact that mutations in different genes can result in similar clinical features.

When a genetic diagnosis is made in an individual, it raises several different types of concerns, not only for the affected individual, but for the relatives as well. For example, if a child is diagnosed with a genetic disorder, the parents often become concerned about the possibility of having another affected child. The siblings and other relatives might wonder about the implications of the genetic diagnosis for their own health or reproductive futures. In late onset disorders, the children or grandchildren of an affected individual may seek information about their risk for developing the condition. For certain types of conditions, like familial cancers, testing may become available that will determine if an individual has an increased susceptibility to develop the disorder.

It is the task of the geneticist and genetic counselor to address these questions and concerns. The family is given information about the inheritance pattern, recurrence risk, and risk to other relatives. The options for testing are also considered.

Genetic Testing by DNA Analysis

DNA-based genetic testing can already be used in a variety of genetic conditions to answer some of the questions and concerns raised in genetic counseling. For some conditions, like cystic fibrosis (CF), X-linked muscular dystrophies and the fragile X syndrome, DNA-based testing can be useful in making or confirming a genetic diagnosis. For these and other conditions, DNA testing can be used to identify individuals who are carriers of the gene mutations that cause the conditions and who, therefore, have an increased risk for having affected children.

For some adult onset disorders, like Huntington Disease (HD) and Adult Polycystic Kidney Disease (APKD), DNA-based testing can identify individuals within a family who have most likely inherited the gene mutation and will therefore probably develop the condition later in life.

DNA-based Diagnostic Testing

In certain situations DNA testing can be used to make or confirm a

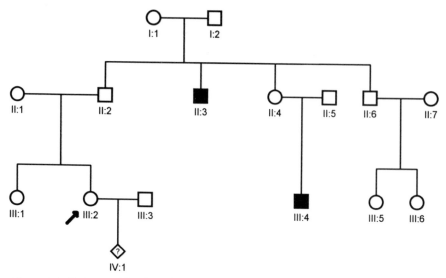

Figure 9.1. Pedigree of family with unclassified mental retardation. Circles are females and squares are males. Filled circles and squares designate affected individuals. See text for details.

genetic diagnosis. A good example is the fragile X syndrome, a relatively common cause of inherited mental retardation in both males and females. The gene that causes the fragile X syndrome (FMR-1) is mapped to chromosome Xq27.3.

Prior to the isolation of the FMR-1 gene in 1991, confirmation of a fragile X syndrome diagnosis was difficult, especially in the absence of a positive family history for the condition. The available cytogenetic testing methods failed to correctly identify all affected individuals. Carrier testing by cytogenetic methods was of limited utility. As described in the previous chapter on trinucleotide repeats, methods now exist that can successfully confirm a mutation in the FMR-1 gene in affected individuals and also identify pre-mutation carriers. The following case example illustrates the usefulness of DNA-based testing for the fragile X syndrome.

Individual III:2 (Figure 9.1) was referred for genetic counseling by her obstetrician because of a positive family history of mental retardation. She is twelve weeks pregnant at the time of referral. As indicated in the pedigree, she had a paternal uncle (II:3) and paternal first cousin (III:4) who had mental retardation of unknown cause. The pedigree was consistent with an X-linked form of mental retardation. Upon the recommendation of the genetic counselor, DNA testing for the fragile X syndrome was performed on the consultant's uncle and confirmed a full mutation in the FMR-1 gene. Additional testing revealed that the consultant's father (II:2) did not carry an FMR-1 mutation, but that his sister (II:4) was a pre-mutation carrier, and her son (III:4) had a full mutation in the FMR-1 gene. The consultant was advised that she faced no increased risk for the fragile X syndrome in her fetus. Genetic counseling and further testing were recommended for other relatives, including the consultant's second uncle (II:6) and his daughters (III:5 and III:6). The presence of a pre-mutation in the FMR-1 gene was confirmed in all three of these individuals.

Several aspects of the genetic counseling process are illustrated by this example. The family history suggested the possibility of X-linked mental retardation. The genetic counselor's recommendation for testing of the affected relative resulted in an accurate genetic diagnosis of the fragile X syndrome in the family. Individuals III:5 and III:6 learned that they were carriers of a pre-mutation in the FMR-1 gene and received genetic counseling about the condition. Each of these women has a 50% risk of transmitting the FMR-1 pre-mutation to their offspring. There is also the risk for expansion of the trinucleotide repeat to a full mutation, resulting in the fragile X syndrome in their offspring.

Had the affected uncle been unavailable or unwilling to be tested, the counselor might have recommended DNA fragile X testing for the consultant alone. Although her testing would have correctly revealed that she is not a carrier of a pre-mutation in the FMR-1 gene, the diagnosis of the fragile X syndrome would have remained undetected in this family. Her cousins (III:5 and III:6) might have given birth to children with the fragile X syndrome without the opportunity of knowing their risks and of the availability of prenatal diagnosis for the fragile X syndrome.

The genetic counselor's recommendations and the cooperative nature of the family resulted in accurate diagnosis for the affected uncle and cousin and in accurate genetic counseling and carrier testing for the cousins. In some families, relatives are unwilling to share information about a suspected or confirmed genetic diagnosis. This can present difficult ethical dilemmas for genetic counselors as will be discussed in the last section of this chapter.

Carrier Testing: Direct Versus Linkage Analysis

Cystic fibrosis (CF) and X-linked muscular dystrophies are two other conditions for which DNA testing is useful in establishing or confirming a clinical diagnosis. In both of these conditions, testing can either be accomplished by direct mutation detection in the gene or by analysis of intragenic or closely linked markers. The advantage of direct mutation analysis is that it can often be used to identify carriers of gene mutations even if the proband is unavailable for testing. The following case illustrates some of the benefits of direct testing and both the benefits and limitations of linkage testing.

Individual III:2 (Figure 9.2) has Duchenne muscular dystrophy (DMD) confirmed by muscle biopsy. DMD is an X-linked recessive condition caused by mutations in the dystrophin gene which is mapped to Xp21. As discussed in the previous chapter on deletion detection, approximately 65% of males affected with DMD have a detectable deletion in their dystrophin gene.

DNA analysis revealed that III:2 had a deletion in exons 46–49 of his dystrophin gene. His parents were referred for genetic counseling. They wanted to have additional children and were concerned about the risk of having another child affected with DMD. Additionally, they were concerned about the implications of the genetic diagnosis for their two other children and for relatives on both sides of the family. Careful review of the family history revealed that there were no other relatives affected with muscular dystrophy.

The genetic counselor explained the X-linked inheritance of DMD,

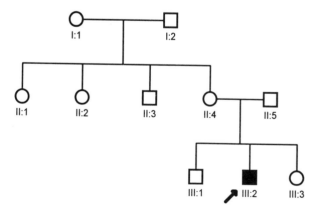

Figure 9.2. Pedigree of family with Duchenne muscular dystrophy. Circles are females and squares are males. Filled circles and squares designate affected individuals. See text for details.

pointing out that the proband's unaffected brother and relatives on the paternal side of the family faced no increased risk for having children affected with DMD.

About one-third of cases of DMD are thought to be due to new mutations in the dystrophin gene. In the absence of a positive family history, it could not be assumed that the proband's mother (II:4) was a carrier of a DMD mutation. DNA analysis was performed and revealed that II:4 did indeed carry a DMD mutation on one of her X chromosomes. The couple was counseled about their one-in-four risk of having a second affected child. They were also told that the proband's sister (III:3) had a 50% chance of being a carrier of a DMD mutation but that accurate carrier testing would be available for her and for the maternal aunts (II:1 and II:2) and other maternally related female relatives who were concerned about their carrier status. The counselor also discussed the availability of accurate prenatal diagnosis by amniocentesis or chorionic villus sampling for any future pregnancy.

Consider the alternative scenario in which III:2 has a confirmed diagnosis of DMD but no detectable mutation in his dystrophin gene. In the absence of a detectable deletion, the mother's carrier status could not be determined by direct mutation analysis.

Some carriers of DMD mutations show a moderate elevation of the muscle enzyme creatine kinase (CK). The proband's mother and sister were tested, and they were both found to have normal CK levels. While this lowered the probability that they are carriers of DMD mutations, since CK is not elevated in all carriers, the normal CK levels could not provide complete reassurance that they were not carriers.

The counselor suggests that the family consider a linkage study which might assist in determining the carrier status of the proband's mother, sister and maternal aunts. The strategy is to compare the marker patterns of the proband and his unaffected brother. If the brothers have identical marker patterns it would imply that the proband's mother is less likely to be a DMD mutation carrier. If the brothers have different marker patterns, the mother's carrier status cannot be confirmed.

The results of the linkage study are presented in Figure 9.3. This example assumes that only one intragenic marker with alternate alleles "1" and "2" was informative. The proband and his unaffected brother (III:1) have different marker patterns. The mother cannot, therefore, be

Figure 9.3. Results of linkage study for Duchenne muscular dystrophy. Marker alleles are indicated as 1 or 2. See text for details.

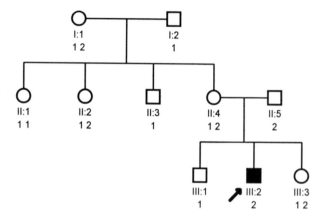

excluded as a possible DMD mutation carrier. The proband's sister (III:3) and one maternal aunt (II:1) have inherited the same marker pattern as the unaffected brother and can, therefore, be reassured that they are unlikely to be DMD mutation carriers. The other maternal aunt (II:2) has inherited the same marker pattern as her affected nephew. Because her sister's carrier status is uncertain, this result is ambiguous. If the proband's mother does carry a DMD mutation, her sister is likely to also be a mutation carrier. If the proband's DMD is the result of a new mutation, the aunt has no increased risk of being a DMD mutation carrier. The linkage study has provided useful information for the proband's sister and for one maternal aunt (II:1), but information that is ambiguous for the proband's mother and for the second maternal aunt (II:2).

This case demonstrates both the potential usefulness and the limitations of linkage testing. Extensive genetic counseling should always be given prior to genetic testing, especially linkage testing. The counseling should include a frank discussion of the limitations of linkage testing. In addition to the possibility of uninformative results, as illustrated in the preceding example, the limitations of linkage testing include the possibility of errors in interpretation due to recombination or because of false assumptions of paternity. Informed consent should be obtained prior to testing. The genetic counseling that precedes testing should provide all the information necessary for people to make informed decisions about whether they wish to proceed with the testing. It is especially important that counselees understand the limitations of genetic testing, and that with a linkage study, errors could be made if the assumptions about paternity are incorrect. Consider again the family presented in Figure 9.3. If individual III:3, the proband's sister, actually had a different father, it would not be possible to exclude her from being a DMD gene carrier without first testing her true biologic father.

Kinship-based Carrier Testing by Direct Mutation Analysis

Cystic fibrosis (CF) is another disorder for which direct mutation detection is available. As described in the previous chapter on cystic fibrosis, over 350 mutations in the CFTR gene have been identified. One mutation (delta F508) accounts for approximately 70% of all CF mutations. Combined with five additional CFTR mutations, the carrier detection rate in the American population of European ancestry is approximately 85%. The carrier frequency and the carrier detection rate varies in different ethnic populations.

Family, or kinship-based, carrier testing for CF is highly accurate, especially if the CFTR mutation status of the affected individual can be identified. Consider the following example. Individual III:4 (Figure 9.4) is referred for genetic counseling and CF carrier testing because his paternal first cousin (III:2) has CF. The condition is inherited according to an autosomal recessive pattern. The parents of an affected individual are both assumed to be carriers of a CFTR mutation. Other relatives have an increased risk of being a CFTR mutation carrier as shown in Table 9.1. Based on the family history, individual III:4 has a 1 in 4

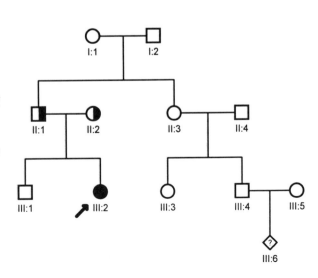

Figure 9.4. Pedigree of family with cystic fibrosis. Circles are females and squares are males. Filled circles and squares designate affected individuals. Mutation carriers are indicated by half-filled circles or squares. See text for details.

Table 9.1. Familial Carrier Risks

Relationship to person with CF	Risk of being a CF mutation carrier
parent	100% (1 in 2 chance)
brother, sister	67% (2 in 3 chance)
aunt, uncle	50% (1 in 2 chance)
half-brother or sister	50% (1 in 2 chance)
first cousin	25% (1 in 4 chance)
no family history of CF	5% (1 in 20 chance)

chance of being a CFTR mutation carrier. The affected cousin has not had CFTR mutation analysis, and the genetic counselor recommends that this be done. The proband (III:2) refuses to be tested. She is upset that her cousin is seeking genetic testing for the purpose of preventing the birth of a child with CF.

CF carrier testing by DNA analysis is performed for III:4. His DNA is screened for the delta F508 mutation and five additional mutations, with a carrier detection rate of approximately 85%. No CFTR mutation is identified in III:4. Using Bayesian analysis, a method commonly used in clinical genetics to assess the relative probability of two alternative possibilities, the genetic counselor calculates III:4's chance of being a CFTR mutation carrier based on family history and the negative CFTR mutation analysis (Figure 9.5). The consultant's final probability of being a CFTR mutation carrier is 4.8% (1 in 21 chances); this is approximately the same as the carrier frequency in the general population (1 in 20 to 1 in 25).

Now consider the alternative scenario in which the proband (III:2) submits to CFTR mutation analysis which confirms that her CF is due to a combination of the delta F508 mutation and a less common mutation R117H. Testing of the proband's parents reveals that she has inherited the delta F508 mutation from her mother (II:2) and the R117H mutation from her father (II:1). The consultant's CFTR mutation analysis did not include the R117H mutation. The consultant is re-tested for this mutation and is identified as a carrier. In this example, accurate carrier testing for the consultant was dependent on knowing the mutation status of the affected relative. Without this information, the consultant might have been falsely reassured by the negative results of the common mutation screen.

	Carrier	Non-carrier
Prior Probability (from pedigree)	1/4 (0.25)	3/4 (0.75)
Conditional Probability (negative mutation analysis; 85% detection)	15/100 (0.15)	1
Joint Probability	(0.25) (0.15) = 0.0375	(0.75) (1) = 0.75
Posterior Probability	0.0375	0.075
	(0.0375) + (0.75)	(0.75) + (0.375)
	0.0476 (4.8%)	0.9523 (95.2%)

Figure 9.5. Bayesian analysis for cystic fibrosis. See text for details.

General Population Carrier Screening: Cystic Fibrosis

For someone with no family history of CF, carrier testing is limited to analysis of the common CFTR mutations. General population carrier testing for CF is controversial. Many concerns about the appropriateness of this testing center on the fact that the currently available testing methods will fail to detect 10–15% of carriers. General population carrier screening for CF will require pre-screening education and counseling that emphasizes the test limitations.

In Americans of European descent, the carrier frequency is about 1 in 25 chances. A couple with no family history of CF, therefore, faces about a 1 in 2500 risk of having a child affected with CF ($1/25 \times 1/25 \times 1/4$). There are no known health risks associated with being a CFTR mutation carrier. Such individuals are, however, at increased risk for having an affected child if their partner is also a carrier. Many people who seek testing or are referred for testing will do so because of concerns about the reproductive risks associated with being a carrier. There is particular concern about the situation where one member of a couple is identified as a definite carrier, and the other has a negative mutation analysis. If the test is 85% sensitive, and the prior risk of being a carrier is 1 in 25 chances, the chance that a person who tests negative is still a carrier is about 1 in 166 chances (0.6%). A couple in which one partner is a definite carrier and one has a negative mutation analysis faces a risk of about 1 in 664 chances ($1 \times 1/166 \times 1/4$). With a test sensitivity of 85%, some at-risk couples in which both partners are carriers would be missed by the screening. Most couples in which one partner tests positive and one tests negative are not at risk, yet some of these couples might unnecessarily avoid reproduction.

The prospect of general population carrier screening for CF has raised several questions and issues regarding the impact of this testing and genetic testing in general on society. These are explored in the recent Office of Technology Assessment report on CF Carrier Screening (U.S. Congress, Office of Technology Assessment 1992). The report identifies eight main factors that affect or will affect CF carrier screening in the general population: genetic services delivery and customs of care, public education, professional capacity, financing, stigmatization and discrimination issues, quality assurance of clinical laboratories and DNA test kits, automation, cost, and cost effectiveness. As the report points out, most of these issues extend beyond CF to global concerns about future genetic tests.

Presymptomatic Testing: Huntington Disease

For adult-onset genetic disorders, DNA-based testing might be used to identify individuals who carry a gene mutation years before the development of symptoms. Testing for Huntington Disease is an example of presymptomatic genetic testing that is already available. Huntington Disease (HD) is an autosomal dominant progressive disease of the central nervous system characterized by involuntary movements (chorea), cognitive impairment and behavioral changes. Currently, there is no cure for HD. The condition can affect both males and females, and the usual age of onset is between 30 and 50 years. Each child of an affected person has a 50% risk of inheriting the HD gene mutation. Virtually all individuals who inherit the HD gene mutation will develop symptoms.

The HD gene was mapped to chromosome 4p in 1983. Until recently only indirect testing by analysis of closely linked DNA markers was available. The HD gene was recently isolated (Huntington Disease Collaborative Research Group 1993). Mutations in the gene involve expansions of CAG trinucleotide repeats, similar to the situation in the fragile X syndrome.

Offspring of individuals with HD face a 50% risk of developing the disorder. Because of the late onset of the disorder, these individuals are

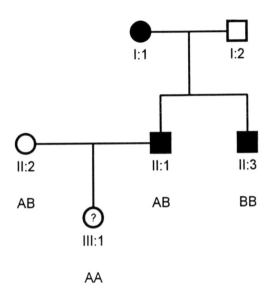

Figure 9.6. Pedigree of family with Huntington disease and informative linkage analysis. Circles are females and squares are males. Filled circles and squares designate affected individuals. See text for details.

in the position of having to make life decisions regarding education, career, marriage and reproduction without knowing if they will develop the disease or have the potential for passing it on to their children.

Presymptomatic testing provides at-risk individuals with the opportunity to learn whether or not they have inherited the HD gene mutation from their affected parent. The testing is usually performed according to a protocol that provides considerable counseling and support before and after testing.

Until recently, this testing could only be accomplished by linkage analysis. Linkage analysis requires the cooperation of relatives and has the potential for uninformative results. Figure 9.6 gives an examples of informative linkage analysis for HD. In this family, the "B" marker pattern is segregating with the HD gene mutation. Individual III:1 has not inherited the "B" marker from her affected father.

Figure 9.7 gives an example of uninformative linkage analysis for HD. The "B" marker pattern is segregating with the HD gene mutation in this family. It cannot, however, be determined if III:1 inherited the "B" marker from his affected mother or his unaffected father.

There is also the potential for errors in prediction of a person's mutation status due to recombination between the gene mutation and the markers that are analyzed. Some individuals might be denied the opportunity to be tested because of lack of cooperation from both affected and unaffected relatives. Consider the situation in Figure 9.8. Individual III:1 wants testing to determine if she has inherited the HD gene mutation. Her paternal grandmother (I:1) who was affected with HD is deceased. Her father (II:2) is 44 years old and has no obvious symptoms of HD. His sister (II:1) is affected, with onset of symptoms at age 50 years. Suppose that III:1's father refuses to participate in testing because he does not wish to know if he inherited the HD gene mutation from his affected mother. His sister (II:1) agrees to be tested to help her niece. The results are shown in Figure 9.8. The testing is uninformative. Although the HD gene mutation can be assumed to be associated with the "B" marker pattern, it can not be determined if III:1 inherited the "B" marker from her unaffected mother or her potentially affected father. If her mother agrees to testing (Figure 9.9), the situation changes. It can now be determined that III:1 inherited the "B" marker pattern from her mother and is, therefore, unlikely to be affected.

The isolation of the HD gene and the availability of direct mutation testing will alleviate some of the former difficulties with pre-symptomatic testing for HD by linkage analysis. Experience with presymptomatic testing for HD has underscored the need for this testing to be done following a strict protocol of pre-test assessment and counseling and post-test counseling and support. While the collective experience of the existing testing programs has provided much insight into the psychosocial ramifications of presymptomatic testing for HD, there is still a need for additional research to evaluate the psychological and social consequences of presymptomatic testing, not only for HD, but for other adult onset disorders (Chapman 1992)

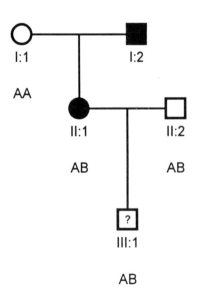

Figure 9.7. Pedigree of family with Huntington disease and uninformative linkage analysis. Circles are females and squares are males. Filled circles and squares designate affected individuals. See text for details.

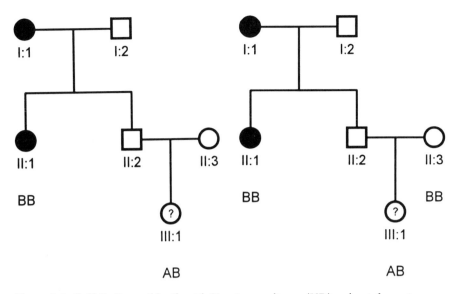

Figure 9.8. (left) Pedigree of family with Huntington disease (HD) and uninformative linkage analysis. Circles are females and squares are males. Filled circles and squares designate affected individuals. The B allele is segregating with HD mutation in this family. It canot be determined if III:1 inherited the B allele from her unaffected mother or her potentially affected father. See text for details.
Figure 9.9. (right) Same family as in Figure 9.8. III:1 inherited her B marker allele from her mother and is, therefore, unlikely to be affected with Huntington disease.

Susceptibility Testing: Familial Breast Cancer

DNA-based testing might also be used to identify individuals who carry mutations in genes that increase their susceptibility for developing certain health problems. Susceptibility testing for familial breast and ovarian cancer is one example. Using linkage analysis in families with familial breast and/or ovarian cancer researchers have localized a breast cancer susceptibility gene (BRCA1) to chromosome 17q12-21 (King *et al.* 1993). It is estimated that 45% of families with autosomal dominant breast cancer and 80% of families with autosomal dominant ovarian cancer show linkage to DNA markers on 17q. Female carriers of BRCA1 mutations are thought to have an 85% lifetime risk of developing breast cancer, with more than 50% of the cancers occurring before age 50. The risk for developing ovarian cancer is also increased in these women, but the magnitude of the risk is unknown. It is unclear if male carriers of BRCA1 gene mutations face any increased risk for cancer.

These men can, however, transmit the BRCA1 mutation to their daughters. There is also evidence that somatically acquired mutations in the BRCA1 gene might play a role in sporadic (non-familial) breast cancer.

Susceptibility testing for BRCA1 germline mutations is not yet clinically available, and the experience in counseling individuals who have inherited BRCA1 gene mutations has been limited to several research families. Genetic counseling about susceptibility testing raises some unique issues (Biesecker *et al.* 1993). Potential risks associated with susceptibility testing for breast cancer include the possibility of jeopardizing insurance coverage. Women who are tested may suffer psychologically as a consequence of the testing. They may experience loss of self esteem, and depression or despair. One potential benefit of such testing is the reduction of anxiety for those identified as non-carriers. For those identified as carriers, increased surveillance might lead to a decrease in morbidity and mortality.

Once the BRCA1 gene is isolated and direct mutation analysis is possible, there may be a demand for population-based screening. A multidisciplinary approach that includes genetic counseling, medical management and ongoing emotional support for mutation carriers will be important to the success of susceptibility testing both within high risk families and in the general population.

Social and Ethical Issues

Genetic Discrimination

The increased availability of genetic testing of all types has raised concerns about possible misuse of genetic information. Of particular concern is the possibility that an individual will suffer social, employment or insurance discrimination based on their genetic status. The term genetic discrimination is used to describe discrimination against an individual or against members of that person's family based solely on real or perceived differences from the "normal" genome of that individual. Billings and colleagues surveyed over 1000 genetics professionals and members of genetic support groups to determine whether incidents which reflect genetic discrimination are occurring in the workplace, in access to social services, in insurance underwriting and in delivery of health care. The survey revealed thirty-two incidents of insurance dis-crimination, seven incidents involving employment discrimination and two incidents involving applications for adoption (Billings *et al.* 1992). The reported examples of discrimination were varied. Some individu-als were denied health insurance even though they had treatable condi-tions. Some carriers of autosomal recessive conditions were denied health insurance despite that fact that there are no health risks associat-ed with carrier status. Many genetic conditions exhibit considerable variability in expression. The survey revealed examples of individuals with very mild expressions of specific disorders who were denied employment, medical and automobile insurance based on their diagnos-tic label alone without regard to their actual health status. There were two examples of individuals, both of whom had a family history of Huntington Disease, whose adoption applications were denied. In one example, a couple with one child with cystic fibrosis had prenatal diag-nosis during a subsequent pregnancy that revealed an affected fetus. They were informed that they would be denied coverage for pregnancy and pediatric care if they continued the pregnancy. The couple threat-ened legal action, and the insurance company reversed the decision. Although this particular study did not attempt to provide quantitative data on the incidence of discriminatory practices, it provided several examples of individuals or families who faced social, employment or insurance discrimination due to their genetic status without regard to their actual health status.

In the United States, the Americans with Disabilities Act (ADA) of 1990 provides some protection against employment and other forms of social discrimination based on genetic status (Alper and Natowicz 1993). The ADA does not provide protection from insurance discrimi-nation based on genetic status. Some people may choose to avoid genet-ic testing because of concerns about possible stigmatization or discrimi-nation.

Population Screening

The increased ability to identify carriers of recessive conditions and individuals who carry gene mutations that predispose them to common health problems is likely to result in more population-based screening programs. Some forms of population-based screening already exist, like newborn screening for phenylketonuria (PKU) and other disorders. The rationale for newborn screening is the identification of infants with genetic disorders who will benefit from presymptomatic treatment. The same principle might apply to screening adults for gene mutations that predispose to common health problems, like heart disease or cancer, to identify individuals who might benefit from health measures for early identification and/or treatment of these disorders. As with newborn screening, it will be necessary to establish the efficacy of the proposed treatment or other interventions before screening programs are instituted. Careful attention also needs to be given to the possibility of adverse psychosocial outcomes to such screening, including the potential for stigmatization or discrimination against gene mutation carriers.

Sickle cell screening in the African-American population and Tay Sachs screening for individuals of Ashkenazi Jewish descent are examples of screening programs to identify individuals who carry recessive genes and who, therefore, have an increased risk for having children affected with these conditions. The experience with these types of screening programs has underscored the need for such programs to have clear goals and objectives. For example, the early sickle cell testing programs frequently did not distinguish the goals for individuals to make informed reproductive choices from the public health goals of reducing the incidence of sickle cell disease. The need for effective education to allow people to make informed decisions about testing and to avoid confusion and inappropriate decisions based on the test results is crucial to the success of these types of screening programs. As with pre-symptomatic testing programs, carrier screening programs need to be planned with careful consideration of the potential for any misuse of genetic information that could lead to discrimination or stigmatization (Fost 1992). Consensus exists that genetic screening programs should be voluntary and that test results should not be released to third parties without the consent of the person who is tested (Wertz and Fletcher 1989).

Ethical Issues: Autonomy, Privacy and Confidentiality

Many ethical concerns regarding genetic counseling and testing center on issues of autonomy, privacy and confidentiality. The tradition of non-directive counseling in genetics recognized the rights of individuals to make informed, independent decisions, free of coercion. Clients are given all the facts necessary to clarify their alternatives and anticipate the consequences of their decisions (NSGC Code of Ethics 1992). Consensus exists that all pertinent information, including ambiguous or conflicting test results, should be shared with clients (Wertz *et al.* 1990). While some would argue that completely value-free counseling is impossible to achieve or that in some circumstances it is appropriate for the geneticists to make their opinions known to clients, the value of protecting client autonomy in decisions regarding genetic testing and reproduction is not seriously questioned.

There is also consensus that individuals have a right to privacy in regard to their own genetic information and that diagnostic information or test results should not be disclosed to third parties without the specific consent of an individual. Questions of confidentiality take on a unique importance in genetic counseling, however, when one considers the sharing of information about a genetic diagnosis *within* a family. Every individual is part of a more extensive family unit. Genetic data, therefore, have something to say about all those belonging to a family. Identifying other relatives who might be at risk for developing a genetic condition or for carrying a particular gene mutation and are thereby at increased risk for having affected children is an important component of genetic counseling. Clients are encouraged to share information about the genetic diagnosis with at-risk relatives. If clients fail or refuse to do this, the genetic counselor faces the difficult dilemma of balancing the client's right to confidentiality and the relatives' rights to be informed of potential risks. There is general agreement that genetic counselors have an obligation to educate clients about the potential impact of the client's genetic status on their relatives. Furthermore, the genetic counselor can help clients to recognize that the clients themselves may have a moral obligation to inform their relatives of potential risks. It is less clear under what circumstances a genetic counselor might be morally justified in breaking a client's confidentiality and directly contacting a relative to inform them of a potential risk. Some guidelines are provided by the President's Commission for the Study of Ethical Problems in Medicine and Biomedical and Behavioral Research (1983), which states that confidentiality may be breached in exceptional circumstances when the following conditions are met: 1) reasonable efforts to elicit voluntary consent to disclosure have failed; 2) there is a high probability both that harm will occur if the disclosed information is withheld and that the disclosed information will actually be used to avoid harm; 3) the harm that identifiable individuals would suffer would be serious; and 4) appropriate precautions are taken to ensure that only the genetic information needed for diagnosis and/or treatment of the disorder in question is disclosed.

Testing by linkage analysis presents a special situation. Samples from several relatives may be required in order to complete the study. In the course of the study, information about the genetic status of a relative who has not specifically requested testing might be revealed. For example, in the course of linkage analysis for a recessive condition, the carrier status of the proband's grandparents might be revealed. The grandparents may have submitted samples to assist in carrier testing for their children but may not wish to know their own carrier status. Similarly, in predictive testing for late onset disorders, the genetic status of individuals who contributed samples for the linkage study might be revealed. Informed consent should be obtained from each person who is tested as part of a linkage study. Agreement should be reached ahead of time about disclosure of test results, and all those who are to receive results should have appropriate counseling. Furthermore, the laboratory reports that become part of a client's medical record should only contain data about the client's genetic status and not the status of relatives.

Most problems concerning confidentiality can be avoided by discussions prior to evaluation and/or testing about the policies with regard to identification of at-risk relatives and disclosure of test results.

Summary

The counseling, social and ethical issues that surround genetic testing have been reviewed in this chapter. Genetic counseling is an important component of any genetic testing. DNA-based tests can be used to make or confirm a genetic diagnosis and to answer various questions raised in genetic counseling.

The current practice of genetic counseling protects the autonomy of individuals and recognizes the rights of individuals to maintain control over access of their own genetic information. Steps to assure that these rights are protected in the future need to be taken.

The increased availability of DNA-based genetic tests is likely to lead to the development of widespread genetic screening. Carrier screening, presymptomatic screening and susceptibility testing for genetic conditions need to be approached carefully so that people will not be unfairly discriminated against because of their genetic status. Effective public education about genetic disorders and genetic testing can help assure that individuals maintain their right to make informed autonomous decisions about genetic testing and reproduction.

References

Ad hoc Committee on Genetic Counseling. Genetic counseling. 1975. *Am J Hum Genet 27*: 240-242.

Biesecker, B., Boehnke, M., Calzone, K., *et al.* 1993. Genetic counseling for families with inherited susceptibility to breast and ovarian cancer. *JAMA 269(15):* 1970-1974.

Chapman, M.A. 1992. Canadian experience with predictive testing for Huntington Disease: Lessons for genetic testing centers and policy makers. *Am J Med Genet 42:* 491-491.

Fost, N. 1992. Ethical issues in genetics. *Ped Clinics of N Am 39(1):* 79-89.

Huntington Disease Collaborative Research Group. 1993. A novel gene containing a trinucleotide repeat that is expanded and unstable on Huntington's disease chromosomes. *Cell 72:* 971-983.

King, M-C., Rowell, S., Love, S. 1993. Inherited breast and ovarian cancer. *JAMA 269 (15):* 1975-1980.

National Society of Genetic Counselors Code of Ethics. 1992. *J Genet Counsel 1(1)*:41-43.

President's Commission for the Study of Ethical Problems in Medicine and Biomedical and Behavioral Research. 1983. *Screening and Counseling for Genetic Conditions.* US Government Printing Office, Washington, D.C.

U.S. Congress, Office of Technology Assessment. 1992. *Cystic Fibrosis and DNA Tests: Implication of Carrier Screening.* U.S. Government Printing Office, Washington, D.C..

Wertz, D., Fletcher, J. 1989. An international survey of attitudes of medical geneticist towards mass screening and access to results. *Public Health Reports 104(1):* 35-44.

CHAPTER 10

Molecular Microbiology

Roy L. Hopfer, Ph.D.

Introduction

Molecular approaches have great potential for use in diagnosis of infectious diseases and for identification of isolated microorganisms, in addition to their application to molecular pathology and to the detection of genetic disorders. Although molecular microbiology is a relatively new scientific discipline, considerable progress has been made during the past five years. As an example of the growth of molecular techniques in clinical microbiology, the *Journal of Clinical Microbiology* had no index references for the polymerase chain reaction (PCR) as recently as 1988; however, more than 100 articles in that same journal were cross-referenced to PCR in the index of volume 30 (1992). In fact, a textbook has been recently published entitled *Diagnostic Molecular Microbiology* (Persing et al. 1993). This book contains 66 PCR protocols that can be used to detect nucleic acids from fifteen bacteria, sixteen viruses, two fungi, and six parasites directly in clinical specimens. Protocols for the PCR detection of eight toxin coding genes and seven antimicrobial resistance genes are also included in the text.

In the UNC Hospitals at the present time, most probes and PCR primers are prepared "in-house," although some commercial systems are available, especially for probe-based identification of isolated organisms. There are numerous sources of basic PCR reagents readily available, such as polymerases, buffers, and additives for increasing the performance of PCR, such as Ampliwax (Perkin-Elmer, Norwalk, CT). For molecular microbiology applications, most commercially available probes are approved for use on colonies of isolated organisms following their culture from clinical specimens, rather than being approved for direct detection in the clinical specimen. Although such probes are very useful in terms of work-flow and finalizing a laboratory report, they are of less value for making patient management decisions on a timely basis.

It is relatively easy to obtain reliable results with probes applied to isolated colonies which have amplified their copy number during growth and colony formation. However, probes directed against genomic fragments present in multiple copies in each organism have proved too insensitive for direct detection in clinical specimens in the majority of systems tested. Similarly, it is relatively easy to amplify genomic fragments from heavy cell suspensions where the efficiency of cell breakage need not be particularly high. However, in clinical specimens where there may be only one or two organisms present, one must have a very high efficiency of cell breakage and subsequent release of DNA into the PCR reaction mix. High efficiency of cell breakage is easily accomplished for viruses and most bacteria, but efficient cell breakage has proved more difficult for mycobacteria, fungi and parasites. These organisms have thick, chemically complex cell walls that are generally resistant to heat, detergents and acid/alkali treatment.

Another potential problem associated with these molecular approaches, particularly PCR, is the possibility that nucleic acids will be detected from dead organisms present in clinical specimens following appropriate antibiotic treatment. This could provide misleading information to the clinician and could be particularly problematic in following treatment of such diseases as tuberculosis. The detection of amplifiable DNA (or RNA) in specimens from such patients makes the evaluation of whether the patient is infectious or not infectious to others very difficult.

Regardless of all the problems associated with these molecular techniques, their potential for providing an early diagnosis, and thereby more rapid initiation of effective treatment, is virtually unlimited. Considerable progress has already been made, and the worldwide research efforts addressing these problems will surely change the currently used methods for assessment of infection in the clinical (molecular) microbiology laboratory. This chapter is limited to discussion of molecular techniques as they apply to two major groups of microorganisms, the mycobacteria and the fungi.

Table 10.1. Examples of commercially available probes.

Type of assay	Organism	Commercial Source
Direct detection in clinical samples	**Bacteria**	
	Chlamidia trachomatis	GenProbe
	Gardnerella vaginalis	MicroProbe
	Group A streptococci	GenProbe
	Legionella pneumophilia	GenProbe
	Neisseria gonorrhoeae	GenProbe
	Parasites	
	Trichomonas vaginalis	MicroProbe
	Viruses	
	Human papillomavirus	Digene
Culture confirmation assays	**Bacteria**	
	Campylobacters	GenProbe
	Enterococci	GenProbe
	Group B streptococci	GenProbe
	Haemophilus influenzae	GenProbe
	Listeria monocytogenes	GenProbe
	Mycobacterium tuberculosis complex	GenProbe
	Mycobacterium avium	GenProbe
	Mycobacterium avium complex	GenProbe
	Mycobacterium intracellulare	GenProbe
	Mycobacterium gordonae	GenProbe
	Neisseria gonorrhoeae	GenProbe
	Fungi	
	Histoplasma capsulatum	GenProbe
	Blastomyces dermatitiditis	GenProbe
	Coccidioides immitis	GenProbe
	Cryptococcus neoformans	GenProbe
	Virus	
	Human papillomavirus	Digene

Adapted with permission from Tenover and Unger 1993.

Probes

Although DNA or RNA probes have been used in research microbiology laboratories since the early 1980s (Moseley *et al.* 1980), the commercial availability of probes for use in the clinical microbiology laboratory has been and remains limited. Three commercial companies offer a total of seven probes for detection of specific nucleic acid fragments from seven organisms in clinical specimens (Tenor and Unger 1993). Also offered are 16 probes for culture confirmation of organisms after their isolation (Table 10.1). In addition, commercial systems using DNA or RNA amplification methods directly on clinical specimens followed by probe analysis are currently being evaluated for the major species of mycobacteria. The interest in automation and commercialization of these and other probes is, however, at a very high level in the industrial community, which will lead to major changes in technical approaches, work flow and time needed for reporting test results to the clinicians.

Nucleic acid-based probes are ideally suited for 1) identification of slow-growing organisms, such as *Mycobacterium tuberculosis*, 2) identification of organisms that are difficult to characterize by classic biochemical analyses, such as *Legionella* species, 3) identification of organisms that have multiple morphotypes, such as members of the black pigmented fungi, 4) determining genotypic relatedness of multiple strains of the same species in epidemiologic investigations, and 5) detection of gene fragments that code for antibiotic resistance. The utility of probes is further enhanced when they are used in conjunction with nucleic acid amplification techniques, such as the polymerase chain reaction, or with restriction enzyme digestion and the resultant restriction fragment length polymorphism (RFLP) analysis, or with dot-blotting hybridization methods.

The only probes used in our laboratory are the mycobacterial culture confirmation probes supplied by GenProbe (San Diego, CA). Since the probes work best if sufficient growth is present to provide visible colonies on solid media, we have both colony morphology and growth rate to assist us in deciding whether to test with either the M. *tuberculosis* or the M. *avium* complex probe. We use the probes to confirm those suspicions, and we finalize the report based on the probe result. This greatly decreases the time needed to identify the organism and can occasionally provide unexpected information, causing direct patient care benefit. For instance, knowing that the isolate is a "mycobacterium other than tuberculosis" (MOTT) allows patients to be removed from isolation. Being removed from isolation saves considerable money for both the patient and the hospital. Direct patient care is also more easily accommodated when the patient is removed from respiratory isolation precautions. The biggest benefit to the laboratory is the tremendous reduction in time and effort ordinarily needed for bio-chemical identification of the mycobacteria.

GenProbe markets specific probes for both M. *avium* and M. *intracellu-lare*. We have elected to use the M. *avium* complex (MAC) probe rather than both the M. *avium* and M. *intracellulare* individual probes. Our clinicians do not feel that speciation beyond the MAC designation offers any direct patient benefit, and it allows us to reduce expenses associated with use of two probes rather than a single probe.

GenProbe is currently evaluating a four-hour amplification probe assay for direct use on clinical specimens (Curry *et al.* 1993; Della-Latta 1993). The reported sensitivity and specificity ranged from 82–98% and 99%–100%, respectively. The major problems associated with the amplification probe system is the interpretation of positive results following antimycobacterial therapy. It is desirable to discharge individuals from the hospital as soon as they become non-infectious. The amplification-probe system may produce a positive result for a certain time following antimicrobial killing of the organism, thereby leading to an extended hospital stay. One would not want to release a potentially infectious immunocompromised patient, such as an HIV-positive patient, to a community housing facility, such as hospice, where other immunosuppressed high-risk patients reside. Correlation of the results from these amplification-based probe systems with clinical findings must be established in order to assess the predictive value of a positive result. The major advantage of these amplification-probe systems is the reduction in the time required for detection and identification of M. *tuberculosis* from the current 2–3 weeks typically required to 24 hours or less.

GenProbe also has commercially available probes for *Cryptococcus neoformans*, *Histoplasma capsulatum*, *Blastomyces dermatitidis* and *Coccidioides immitis*. We do not use any of these probes in our mycology laboratory. *Cryptococcus neoformans* is generally easy to grow and identify, and most patients are initially diagnosed using cryptococcal latex agglutination tests for capsular polysaccharide. We are not in endemic areas of either coccidioidomycosis or histoplasmosis, and we have only two or three isolates of B. *dermatitidis* each year. Therefore, it is not economically feasible for our laboratory to keep probe kits available for these organisms. In addition, most of our patients have tissue histopathology findings consistent with the pathogen isolated. Because of the histologic findings, we rarely need a quick or rapid identification of a fungal isolate that is morphologically typical of the suspected pathogen. Most strains of these are rather easy to identify. On occasion we have had poorly sporulating or difficult to convert (mold to yeast phase) isolates and, in those instances, we have sent the isolates to nearby commercial laboratories that use the GenProbe fungal probes to confirm the identification.

PCR Methods

Mycobacteria

There are at least 12 primer systems that have been published for the detection of M. *tuberculosis* and/or related *Mycobacterium* sp. GenProbe, Roche Molecular Systems, Inc. (Branchburg, NJ), and GeneTrak Systems (Framingham, MA) are all actively pursing development of various amplification systems (PCR and Q Beta Replicase) for detection of mycobacteria in clinical specimens, but none are commercially available at the present time. We currently use two of the reported PCR systems (Eisenach *et al.* 1990; Pao *et al.* 1990) in our laboratory. The Eisenach primer set amplifies a multiple repeat segment which is specific for M. *tuberculosis* and does not amplify genomic segments from other mycobacteria. The Pao primer set, which amplifies a segment from the 65 kilodalton antigen, can detect

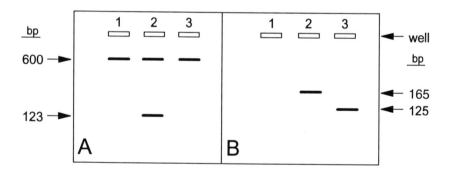

Figure 10.1. Diagrammatic representation of expected PCR results using Eisenach primer set (panel A) and Pao primer set (panel B). Band of 600 bp is due to addition of Eisenach internal control to all test specimens. Specimens from: Uninfected patient (Lane 1); patient with tuberculosis (Lane 2); patient with M. *avium* infection (Lane 3).

all mycobacteria. However, the amplified base pair (bp) size of the non-tuberculosis species is 20–40 bp smaller than the fragment from M. *tuberculosis*. Therefore, the combination of both the Eisenach and Pao primer sets provides confirmation or "insurance" beyond a single primer system. For instance, a patient with tuberculosis (Lane 2) should have a positive PCR result with both the Eisenach and Pao primer sets, a 123 and 165 bp fragment, respectively. In addition, an internal control for the Eisenach primers would give a 600 bp band, which indicates the clinical specimen is not inhibitory to the PCR process (Figure 10.1A). Alternatively, a patient with M. *kansasii* infection or with M. *avium* (Lane 3) infection or colonization would have a positive PCR result of a single 600 bp and a 125–145 bp fragment for the Eisenach (internal control) and the Pao primer sets, respectively (Figure 10.1B).

For this method to have the desired sensitivity, the specimen processing method must include an efficient cell (mycobacterial) breakage or lysis step. Mechanical breakage with glass beads and shaking or sonication, chemical breakage with enzymes and/or strong alkali, and boiling in detergent and/or alkali have all been reported. We have had best results using a combination of boiling (10 min) in the presence of sodium dodecyl sulfate.

All specimen processing through the boiling step is performed in contained systems with tubes being opened only in a biosafety laminar flow hood. Respiratory specimens are digested and decontaminated using standard methods (Roberts et al. 1991) prior to PCR processing. At the present time, the PCR-based processing is too labor intensive for our laboratory to perform on all specimens submitted for mycobacterial culture. Therefore, we use PCR only on smear-positive clinical specimens and on culture fluids from Bactec bottles when they attain a growth index of > 100. Even with these criteria, i.e., when relatively large numbers of mycobacteria are present, we have frequent false negative PCR results. Because of the lack of sensitivity of the PCR to date, we have been using the PCR detection of mycobacteria only as an investigational approach and not as a routine clinical assay.

Fungi

It is easy to justify and understand the desire to have PCR detection methods for some of the slow-growing fungi, such as *Histoplasma capsulatum*, or nonculturable organisms, such as *Pneumocystis carinii*. However, the most frequent fungal isolates in clinical mycology laboratories are *Candida albicans* and related *Candida* species. These organisms generally grow within 24–36 hours and are easily recognized microscopically. Further, these organisms are considered a part of the normal microbiota of the human host, so that their detection by PCR in non-sterile clinical specimens sources would be difficult to interpret. Superficial infections caused by these organisms, such as oral thrush or candidal vaginitis, are easily diagnosed by clinical symptoms and culture. Serious life-threatening candidal infections (candidiasis) are, quite paradoxically, very difficult to diagnose in a timely manner that is beneficial for patient management. Patients with life-threatening disseminated candidiasis almost always have a serious immune deficit due to one of a variety of underlying diseases or therapies. Immune suppression due to malignant processes, cytoreductive antitumor treatment, maintenance therapy to prevent rejection following organ transplantation, and immune suppression caused by infectious agents are frequently associated with disseminated candidiasis. One of the major reasons that diagnosis of disseminated candidiasis is often delayed is because blood cultures are frequently negative. When blood cultures for C. *albicans* are positive in this patient setting, they often become positive just prior to death of the patient. Ideally, one would hope to detect the disseminated infection days (if not weeks) earlier so that appropriate antifungal therapy could be given. Other diagnostic tests, such as assays for the detection of antibody, antigen and specific fungal metabolites, have proved too insensitive to be of reliable clinical value. Finally, serious fungal infections caused by other fungi, such as *Aspergillus fumigatus*, are often equally difficult to diagnose, and the diagnosis is frequently made at post-mortem examination.

Highly immunosuppressed patients can also have life-threatening infections caused by bacteria, viruses and parasites. Therefore, it would be of great clinical value to be able to discern fungal infections from infections caused by these other organisms. In an effort to accomplish that level of discrimination, we elected to use a PCR primer set that recognizes a fragment of the small nuclear ribosomal RNA gene (rDNA). This gene segment has been conserved in fungi throughout evolution. To date, this 310 base pair segment has been detected by PCR amplification in all fungi tested. Of equal importance, this rDNA fragment has not been amplified or detected in any prokaryotic cells.

Amplification and detection of this rDNA fragment in clinical specimens would provide highly specific evidence of the presence of fungi. If the specimen is collected from a normally sterile site, the strong implication would be that the life-threatening infection is due to fungi rather than bacteria, viruses, etc. This critically important information would be used for selection of appropriate antimicrobial therapy.

A few years ago the only appropriate treatment for serious fungal infections was amphotericin B, surgery, or a combination of both. Since the recent introduction of the more effective, less toxic, azole class of antifungals, their use has benefitted the clinician by allowing better selection of different treatments based on the knowledge of the identity of the fungus involved. For instance, an infection caused by a non-septate zygomycete, such as a *Rhizopus* species, generally requires more aggressive surgical debridement than if the infection is caused by a septate organism, such as an *Aspergillus* species. Likewise, patients with disseminated or meningeal cryptococcal infections receive different antifungal therapy than patients with disseminated candidiasis. Because of these concerns, we have developed restriction fragment length polymorphism (RFLP) analyses of the PCR-amplified product to differentiate groups of medically important fungi.

The GenBank data system contains known nucleic acid sequences of a variety of genes that have been cloned from both medically related and non-medically related fungi. Using the GenBank data analysis system, we were able to compare the gene fragment that is amplified using the NS5–NS6 rDNA primers (White *et al.* 1990). There were a total of 18 fungi, primarily yeasts, whose DNA sequence of the NS5–NS6 amplifiable region was known or at least partially known. In addition there were seven other organisms that contained compatible primer sites that could yield PCR-amplifiable material using the NS5–NS6 primer set. These latter organisms are rarely, if ever, associated with clinical disease or found in clinical specimens.

The GenBank data system also provided information on the presence, absence and location of over 180 different restriction enzyme sites. We compared restriction site differences among the 25 organisms to determine which restriction enzymes might be of value for our needs. Since there were only 18 medically related species of fungi available for analysis, identifying RFLPs in the other closely related fungi was achieved by trial and error. We are currently sequencing certain groups of fungi in order to make this task less cumbersome and more predictive, and therefore more reliable.

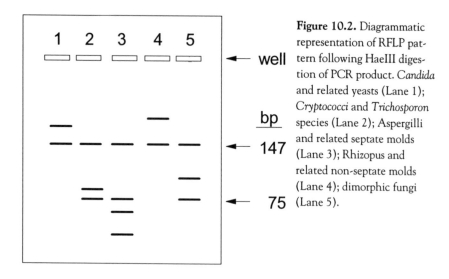

Figure 10.2. Diagrammatic representation of RFLP pattern following HaeIII digestion of PCR product. *Candida* and related yeasts (Lane 1); *Cryptococci* and *Trichosporon* species (Lane 2); Aspergilli and related septate molds (Lane 3); Rhizopus and related non-septate molds (Lane 4); dimorphic fungi (Lane 5).

The enzyme for which there appeared to be the most variety of locations and numbers of restriction sites available in the 18 organisms of interest was HaeIII. *Candida albicans* and other *Candida* species have a HaeIII site at bp 147, giving rise to two fragments (163 and 147 bp in size); *Cryptococcus neoformans* and *Trichosporon beigelii* have two HaeIII sites, giving three fragments (147, 87 and 76 bp in size); *Blastomyces dermatitidis* and *Coccidioides immitis* also have two HaeIII sites, giving three fragments (147, 90 and 73 bp in size); and *Aspergillus fumigatus* contains three HaeIII sites giving four fragments (147, 74, 59, 30 bp in size). There were no sequence data available for the non-septate organisms, such as *Rhizopus* species, but HaeIII digestion suggests that the PCR product obtained in these organisms is approximately 341 bp in length with one restriction site, giving rise to two fragments (147 and 194 bp in size). Therefore, we have shown that the RFLP patterns of the PCR product can be used to sort organisms into each of five taxonomically (and clinically) related groups of medically important fungi (Hopfer et al. 1993) (Figure 10.2). It is, however, unrealistic to expect that other RFLP patterns will not be found since these data are based on such a limited number of organisms. We have recently found that NS6 can misread human DNA and produce a PCR product, especially in high Mg^{++} concentration. We are, therefore, moving NS6 upstream 8 bases wich will remove the "false positive" results with human DNA.

As shown in Table 10.2, other restriction enzymes can be used to further differentiate certain organisms. One enzyme, HincII, appeared to have a restriction site present in the PCR fragment derived from *C. neoformans*, but to have none in any of the other 17 organisms in the database. The HaeIII patterns of *C. neoformans* and *T. beigelii* (a taxonomically related yeast) were identical; however, the treatment of these infections is quite different. Therefore, digestion with HincII was used to analyze the PCR product from these two organisms as well as from four additional saprophytic species of *Cryptococcus*. Since *T. beigelii* and the other *Cryptococcus* species were not in the database, we were not sure if HincII would be helpful. It was of interest that HincII cut *C. neoformans* to give two fragments (247 and 63 bp in size), while *T. beigelii* also had one restriction site (giving fragments 200 and 110 bp in size). The saprophytic cryptoccoci apparently had no HincII restriction sites, which resulted in an uncut band of 310 bp. Therefore, one enzyme (HincII) can be used to readily distinguish *C. neoformans* from other saprophytic *Cryptococcus* species and from *T. beigelii* (Figure 10.3).

Although we are continuing these studies to further delineate the value of this RFLP procedure, we recognize that there are several shortcomings

Table 10.2. Restriction enzymes that can be used to differentiate between specific organisms or groups of organisms following digestion of product of an amplified region of the small nuclear ribosomal RNA gene.

Enzyme	Differentiates
HaeIII	*Candida* and related yeasrs *Cryptococci* and *Trichosporon* *Aspergillus* and related molds *Rhizopus* and related molds Dimorphic fungi
HincII	*Cryptococcus neoformans* *Trichosporon beigelii* Saprophytic cryptococci
HphI	*Pneumocystis carinii*
TfiII	*Sporothrix schenckii*
MslI	*Coccidioides immitis*
HaeII	*Candida lusitaniae*
NcoI	*Candida krusei*
TaqI	*Candida lipolytica*

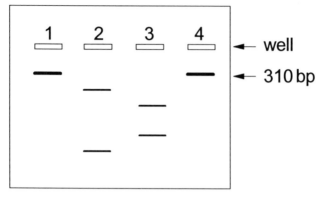

Figure 10.3. Diagrammatic representation of RFLP pattern following HincII digestion of PCR product. *Candida albicans*, or other *Candida* species (Lane 1); *Cryptococcus neoformans* (Lane 2); *Trichosporon beigelii* (Lane 3); and saprophytic *Cryptococcus* species (Lane 4).

or disadvantages. All of the data that we have generated to date is based on characterizing a very small fragment of the rDNA. In order to make the test more valuable, we may have to expand the size of the target DNA well above the current size of 310 bp. We also recognize that restriction enzymes are a helpful research tool, but other methods, such as the use of specific probes, could be used to reduce the time needed to analyze the PCR product. We are presently investigating heteroduplex formation as an alternative to multi-step RFLP analyses. A single base pair difference between the amplified products can be detected by gel electrophoresis by this latter technique. Ideally, heteroduplex analysis could be accomplished in one electrophoresis run without the need to repeat digestions inherent in the RFLP analyses.

Conclusion

The timely detection of mycobacteria and fungi in clinical specimens remains a laboratory dilemma. Because of their slow growth and presumed low numbers, they are ideal candidates for detection by molecular approaches, such as PCR, RFLP and probe technology. To date, the only commercially available kits are nucleic acid-based probes used for confirming the identity of an organism after it has grown. Amplification techniques are hampered by the inability to efficiently break or digest the cell wall, allowing the release of nucleic acids. These technical problems are being investigated in many laboratories, and the use of these techniques on direct clinical specimens should be applicable to clinical laboratories in the near future. These, or similar amplification techniques, will reduce the turn-around time for detecting and identifying these slow-growing organisms from weeks to hours.

References

Curry, J. I., Wolfe, J.M., Moore, D.F. 1993. Detection and identification of *Mycobacterium tuberculosis* directly from induced sputum specimens using amplification of ribosomal RNA. In *Abstracts of the General Meeting of the American Society for Microbiology*. Washington, D.C: American Society for Microbiology, p 176.

Della-Latta, P., Waithe, E., Sorondo, G., Lungu, O., Whittier, S., Silverstein, S. 1993. Rapid detection of *Mycobacterium tuberculosis* directly from sputum by the GenProbe Amplified *Mycobacterium tuberculosis* Test. In *Abstracts of the General Meeting of the American Society for Microbiology*, Washington, D.C.: American Society for Microbiology, p. 177.

Eisenach, K.D., Cave, M.D., Bates, J.H., Crawford, J.T. 1990. Polymerase chain reaction amplification of a repetitive DNA sequence specific for *Mycobacterium tuberculosis. J Inf Dis* 161:977-81.

Hopfer, R.L., Walden, P., Setterquist, S., Highsmith, W.E. 1993. Detection and differentiation of fungi in clinical specimens using polymerase chain reaction (PCR) amplification and restriction analysis. *J Med Vet Mycol* 31: 65-75.

Moseley, S.L., Hug, I., Alim, A.R.M.A., So, M., Samadpour-Motalebi, M., Falkow, S. 1980. Detection of enterotoxigenic *Escherichia coli* by DNA colony hybridization. *J Inf Dis* 142: 892-898.

Pao, C.C., Yen, T.S.B., You, J.B., Maa, J.S., Fiss, E.H., Chang, C.H. 1990. Detection and identification of *Mycobacterium tuberculosis* by DNA amplification. *J Clin Microbiol* 28: 1877-80.

Persing, D.H., Smith T.F., Tenover, F.C., White, T.J. 1993. *Diagnostic Molecular Microbiology*. Washington, D.C.: American Society for Microbiology.

Roberts, G.D., Koneman, E.W., Kim, Y.K. 1991. Mycobacterium. In *Manual of Clinical Microbiology*, edited by Balows, A., Hausler, W.J., Jr., Herrmann, K.L., Eisenberg, H.D., and Shadomy, H.J. Washington, D.C.: American Society for Microbiology, pp 304-39.

Tenover, F.C., Unger, E.R. 1993. Nucleic acid probes for detection and identification of infectious agents. In *Diagnostic Molecular Microbiology*, edited by Persing, D.H., Smith, T.F., Tenover, F.C. and White, T.J. Washington, D.C.: American Society for Microbiology, pp 3-25.

White, T.J., Bruns, T., Lee, S., Taylor, J, 1990. Amplification and direct sequencing of fungal ribosomal RNA genes for phylogenetics. In *PCR Protocols A Guide to Methods and Applications*, edited by Innis, M.A., Gelfand, D.H., Sninsky, J.J., White, T.J. San Diego: Academic Press, Inc. pp 315-22.

CHAPTER 11

Molecular Biological Techniques for the Detection of Human Immunodeficiency Virus Type 1

Susan A. Fiscus and Joseph J. Eron

Introduction

The history of the polymerase chain reaction (PCR) parallels the history of our knowledge of human immunodeficiency virus type 1 (HIV-1) and acquired immunodeficiency syndrome (AIDS). In 1983, while French scientists were reporting the isolation of HIV-1, the etiologic agent of AIDS (Barre-Sinoussi et al. 1983), Dr. Kary Mullis at Perkin-Elmer-Cetus was initiating experiments which led to the invention of the PCR technology (Templeton 1992). In 1985, as the first diagnostic tests for HIV-1 were being approved and put into use, the first published applications of the PCR appeared. In 1987, while zidovudine (ZDV), the first drug to be used against HIV-1, was approved, the first studies of HIV-1 using PCR technology were published (Kwok et al. 1987). And in 1989, as the initial reports of the appearance of HIV-1 isolates resistant to ZDV among treated patients were being published (Larder et al. 1989; Larder and Kemp 1989), PCR was being used in the detection of HIV-1 infection in infants (Rogers et al. 1989) and in individuals who had not yet seroconverted (Loche and Mach 1988; Imagawa et al. 1989).

Through the use of molecular biological techniques, the genome of HIV-1 has been cloned and sequenced, spliced and recombined, deleted and mutated, all in an effort to understand better the pathogenesis of HIV infection and to devise safe and efficacious drugs and vaccines to combat the disease. This chapter will describe some of the uses of PCR technology in the clinical retrovirology laboratory.

The Polymerase Chain Reaction

The PCR is a powerful tool that can be used to amplify virtually any segment of DNA, provided enough sequence information exists to create
oligonucleotide primers for the PCR reaction. The technique is simple
and lends itself to automation, which is important in a clinical setting.

Reagents

Early descriptions of PCR for HIV-1 included labor-intensive procedures for the preparation of either DNA or RNA from peripheral blood
mononuclear cells (PBMC) or plasma. For instance, the PBMC were isolated on Ficoll-Hypaque gradients, washed extensively, treated with detergents and proteases, and finally extracted with organic solvents, such as phe-
nol-chloroform. Plasma had to be rapidly separated from the whole blood
and treated with guanidinium thiocyanate to inhibit ribonucleases prior to
organic solvent extraction (Chomczynski and Sacchi 1987). More recently
these steps have been simplified considerably. For instance, red blood cells
from EDTA or acid-citrate-dextrose anti-coagulated blood can be lysed,
DNA can be extracted from the white blood cells, using proteinase K,
non-ionic detergents and heat, without the need for organic solvents
(Jackson 1993). Dried blood spots have been used successfully as a source
of target DNA, and their use may become important if screening of infants
becomes more widespread. Cassol *et al.* (1992) found that all 19 HIV-1
infected children in their study were positive by PCR using only about
50 ul of blood spotted on a filter paper and stored at -20°C for 10–26
months. These results included samples from six infants who were less
than ten weeks of age.

RNA sequences can also be amplified if they are first converted to complementary DNA by reverse transcription. Typically, avian or murine
reverse transcriptase is used in this step, although recently it has been
found that the DNA polymerase from *Thermus thermophilus* can catalyze
both reverse transcription and DNA polymerization. Plasma appears to
be a better sample for HIV RNA extraction than serum (Katzenstein *et al.*
1992), and sodium citrate or EDTA should be used as anti-coagulants
instead of heparin (Holodniy *et al.* 1991). Guanidinium thiocyanate has frequently been used to inhibit ribonucleases in the preparation of HIV RNA
for PCR (Chomczynski and Sacchi 1987). It appears important to remove

the plasma from the cellular fraction of whole blood as rapidly as possible. Immediate centrifugation in Leukoprep (Becton-Dickinson, Rutherford, NJ) tubes has been found to stabilize the HIV RNA for up to 48 hours (Yen-Lieberman *et al.* 1993). Henrard *et al.* (1992) developed a method which uses microbeads coated with monoclonal antibodies specific for HIV-1 glycoproteins to capture virions prior to reverse transcription and PCR. Once the nucleic acid has been prepared, it can either be stored at -70°C or amplified immediately.

The choice of primers to be used for HIV-1 amplification depends on the specific application. The sensitivity and specificity of a PCR assay can be influenced by the primer pair used, the magnesium chloride concentration in the reaction mixture, and the annealing temperature. For diagnostic purposes, regions with little cross-reactivity with other retroviruses should be selected for amplification. A number of primer pairs and their probes specific for particular regions of HIV-1 have been tested with clinical specimens from North America and Europe, and their sequences have been published (Coutlee *et al.* 1991). Some of these primers may not be as efficient in amplifying HIV-1 sequences from African countries (Candotti *et al.* 1991), although primers designed using North American isolate sequences were successfully used in a study of perinatal HIV-1 infection in Ugandan infants (Jackson *et al.* 1991).

Methods of Detecting PCR Products

Detection of the amplified sequences can be achieved by ethidium bromide staining of the PCR product after electrophoresis in agarose gels if there is sufficient DNA. Ethidium bromide detection usually requires 10 ng of amplified product. The use of two amplification series and nested primers may increase the sensitivity of the assay and result in a product that can be visualized by ethidium bromide staining, although these methods increase the risk of carry-over contamination. Williams *et al.* (1990) investigated vertical transmission using four sets of nested primers and found that all 22 samples from their nine symptomatic infected patients were positive by PCR as detected by ethidium bromide staining, while all 50 samples from 44 apparently uninfected children were negative. Scarlatti *et al.* (1991) reported similar findings.

Until recently, the most common methods of detecting PCR products were by Southern blotting or liquid hybridization with radiolabeled probes

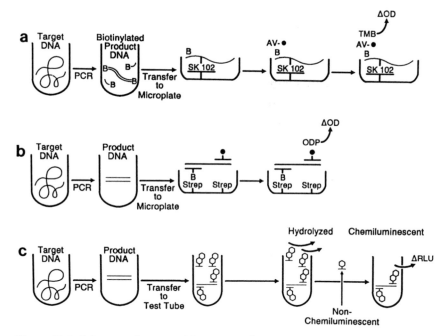

Figure 11.1. Schematic diagrams of three non-radioactive HIV-1 detection assays. (a) Roche HIV-1 PCR assay. (b) Du Pont ELOSA. (c) GenProbe hybridization protection assay. • = horseradish peroxidase; Av = avidin; biotin; TMB = tetramethylbenzidine; OPD = ortho-phenylene diamine; Strep = streptavidin; RLU = relative light units; ⏀ = acridinium ester (not correct chemical formula).

which were specific for the target DNA, followed by autoradiography. This method is sensitive and confirmatory for the size and for specific identification of amplified DNA. However, this method requires the use of radioactive isotopes, is labor-intensive, and its interpretation is somewhat subjective, especially if weakly positive reactions are observed.

More recently non-isotopic systems for detecting HIV-1 specific sequences have become commercially available (Figure 11.1). These include the hybridization protection assay developed by GenProbe Corp. (San Diego, CA), in which a chemiluminescent acridinium ester-labeled probe hybridizes with the amplified HIV-1 *gag* sequences. After the addition of base to the sample, the probe is hydrolyzed and loses its luminescent property unless it is hybridized to the target DNA. The amount of probe:target heteroduplex material is proportional to the intensity of the chemiluminescent signal as measured in a luminometer.

Another non-radioactive detection system developed by Roche Molecular Systems (Nutley, NJ) employs biotinylated primers specific for HIV-1 *gag* sequences and thus can be detected by a simple enzyme-linked absorbance assay. Amplified products are incubated in wells of a 96-well plate that is coated with complementary sequences which capture the specific *gag* sequences. Biotin-labeled captured DNA sequences are detected by the addition of horseradish peroxidase-labeled streptavidin and tetramethylbenzidine substrate.

An enzyme-linked oligonucleotide sandwich assay (ELOSA) has been developed by Du Pont (Billerica, MA). After amplification, the DNA products are added to streptavidin-coated microtiter wells. Biotin-labeled capture oligonucleotide and horseradish peroxidase-labeled reporter oligonucleotide complementary to the same strand of the PCR product are added to each well. If the sample contains the specific HIV-1 sequence, a hybridization complex is formed by the PCR product, the capture probe and the reporter probe, and this complex is attached to the microtiter well by biotin-avidin affinity.

The three commercially available kits (GenProbe, Du Pont, and Roche) were recently evaluated for sensitivity and specificity and compared with a standardized liquid hybridization and autoradiographic protocol (Whetsell *et al.* 1992) (Figure 11.1). Sensitivities and specificities ranged from 96–100% among the four assays.

HIV-1 Quantitation by PCR

Quantitation of nucleic acid by PCR is difficult since the amount of final reaction product depends on the exponential amplification of the initial DNA copy number. Minor differences in amplification efficiency can lead to large and unpredictable differences in the yield of the final product. Factors inherent in the performance of the PCR include sample preparation, the non-linear enzymatic activity of *Taq* polymerase, the structural conformation of the target sequence, the length of the segment to be amplified, and machine performance. Substances which might inhibit PCR, such as hemoglobin and heparin, must be avoided or controlled for. The key factors that affect the quantitative ability of a PCR procedure have been reviewed recently (Ferre 1992; Yang *et al.* 1993). In developing a quantitative PCR assay, every step of the procedure from sample preparation to product detection must be carefully controlled. Internal standards, used in addition to external positive and negative controls, can detect variability in reaction conditions, machine cycling, sample preparation and sample integrity, and may be the most accurate way to use PCR quantitatively.

Quantitative Methods for HIV-1 DNA

Quantitative analysis of specific DNA using an external standard implies that the amount of starting material must be known either by accurate cell counting or by determining the amount of DNA spectrophotometrically by measurement of the absorbance at OD260. Quantitation in these studies is usually accomplished by comparing signals from amplifications of dilutions of the patients' PBMC with an unknown copy number with the signals generated by a standard curve from a chronically infected cell line (Dickover et al. 1990; Eron et al. 1992). Alternatively, an endpoint dilution series of a patient's PBMC can be lysed and subjected to PCR to estimate the HIV-1 DNA copy number per million PBMC (Poznansky et al. 1991).

However, these methods of quantification using an external standard do not really control for differential efficiencies and kinetics of PCR which depend, in part, on the starting amount of the target DNA, the sequence match of the primers, and variable amounts of potential PCR inhibitors which might be present. A third method for the quantitation of HIV DNA involves the use of an internal standard which is largely identical but yet distinguishable from the intended HIV sequence (Telenti et al. 1992). Serial dilutions of known quantities of this competitive template are added to replicate PCR assays containing identical aliquots of the unknown sample to be quantitated. This internal standard template acts as a competitor at all steps in the amplification process. When the amounts of target and competitor DNA are equivalent, equal amounts of each product will be produced. This approach seems to have excellent sensitivity and reproducibility but may be too labor-intensive and expensive for clinical trials, since a minimum of 4–8 amplifications have to be performed on each patient sample at each time point in order to quantitate the amount of HIV DNA in a particular specimen.

Quantitative Methods for HIV-1 RNA

Quantitation of HIV-1 RNA after reverse transcription and PCR requires the inclusion of several standards which serve as internal controls for the

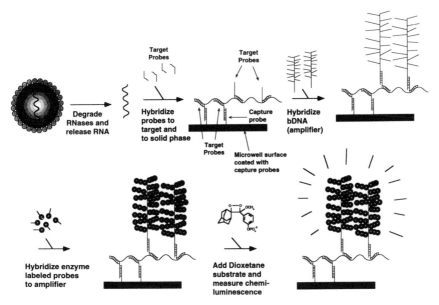

Figure 11.2. Schematic diagram of the Chiron branched DNA assay.

different steps of the procedure. Both the reverse transcription and DNA amplification steps must be monitored. In most cases the plasma has been treated with guanidinium thiocyanate and the RNA extracted with phenol/chloroform prior to alcohol precipitation. A number of different methods have been described both for quantitating the amount of HIV-1 specific RNA and for detecting the product, and some of these assays are being developed commercially.

Chiron Corporation has developed a new signal amplification assay using branched DNA for the quantitation of HIV-1 RNA in plasma (Pachl *et al.* 1993) (Figure 11.2). Virions in plasma are ultracentrifuged, and the pellets resuspended in an extraction buffer which also contains the HIV-1 specific target oligonucleotide probes. After incubation and a centrifugation, the sample is transferred to a microtiter plate coated with capture probes which hybridize to target probes. Branched DNA, also specific for the target probes, is added, followed by enzyme labeled-probes that bind to the branched DNA-RNA-probe complex. Detection is achieved by incubating the complex with dioxetane, a chemiluminescent substrate, and measuring the light emission. If HIV RNA was present in the sample, a sandwich is formed between the capture probe, the HIV RNA, the target probe, the branched DNA, and the enzyme-labeled probe. Quantitation is

A. QC-PCR: PRINCIPLE AND SCHEMATIC ILLUSTRATION

Wild type target sequence template (260 bp product) added, 10 copies constant

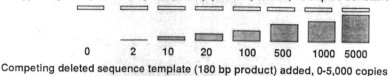

| 0 | 2 | 10 | 20 | 100 | 500 | 1000 | 5000 |

Competing deleted sequence template (180 bp product) added, 0-5,000 copies

B. EtBr STAINED GEL OF QC-PCR ANALYSIS: 10 COPIES HIV

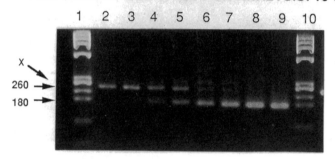

Figure 11.3. Schematic diagram of the quantitative competitive PCR methods. (A) Diagram demonstrating the principle and expected results for quantitative competitive PCR analysis of 10 copies/reaction of wild-type HIV-1 target sequence. (B) Ethidium bromide-stained gel for quantitative competitive PCR analysis of 10 input copies of wild-type HIV-1 *gag* DNA. There is progressive competition between the fixed amount (10 copies) of wild-type per reaction (260 bp band; lanes 2–9) and from 0 to 5000 copies of competitive template per reaction (180 bp band, lanes 2–9). Equivalence point, after correcting for the relative masses of the protein bands, is in lane 4, to which 10 copies were added as a competitor. "X" refers to putative heteroduplex formed near the equivalence point. Lanes 1 and 10 contain size markers.

determined by comparison of the signal generated from a standard curve. The signal is directly proportional to the amount of HIV RNA in the specimen. This assay has recently been shown to be more sensitive than plasma culture (Bobey *et al.* 1993; Cao *et al.* 1993; Dewar *et al.* 1993; Pachl *et al.* 1993; Rasheed *et al.* 1993). In four different studies the sensitivity of the branched DNA assay in HIV-1 seropositive patients ranged from 76–98%, compared with 18–61% sensitivity for plasma cultures. Importantly, the presence of heparin does not affect the branched DNA assay, and so this technique may be useful for determining the effects of anti-retroviral drugs retrospectively in those trials in which only heparinized plasma was stored. However, one milliliter of plasma is necessary for each determination.

More recently HIV-1 RNA has been quantitated with competitive PCR assays (Menzo *et al.* 1992; Scadden *et al.* 1992; Piatak *et al.* 1993) (Figure 11.3). These methods are similar to those described for the competitive DNA PCR assays. A competitive RNA template matched to the target of interest but differing slightly to allow separate detection is used to compete in the reverse transcription and DNA amplification steps, serving as an internal control. However, this method requires numerous (4–8) dilutions of the competitor for each sample tested and may be too labor-intensive and expensive in its present format to be used in clinical trials.

Roche Molecular Systems has modified this technique somewhat to make it more suitable for monitoring clinical trials. The assay uses a noncompetitive internal standard, which is a synthetic RNA transcript containing the identical primer binding sites as HIV-1 and is of the same size. The sequence of the intervening region of the internal standard can be detected separately. Each patient's plasma sample is spiked with 100 copies of the internal control prior to reverse transcription and DNA amplification. A dilution series of the internal control generates a standard curve. Detection is achieved by a modification of Roche Molecular System's previously described enzyme-linked assay to provide a broader dynamic range.

Contamination

The greatest problem with PCR in the clinical lab is false positive reactions that result from contamination. Because of the exquisite sensitivity of PCR, the inadvertent transfer of a tiny amount of target DNA into a neighboring tube can lead to a false positive result. Meticulous care must be taken to avoid contamination. Specific steps which should be used have been described in detail elsewhere (Clewley 1989; Kwok and Higuchi 1989). For instance, amplification reactions should be set up in rooms separated from sample preparation and post-amplification testing. Dedicated supplies, reagents and pipettes should be provided that either never or only come into contact with amplified DNA. Reagents should be aliquoted and used only once. Positive displacement pipettes or aerosol-resistant pipette tips should be used. Aerosols should be carefully avoided and gloves changed frequently. Nested PCR reactions are particularly prone to contamination.

A new technique that greatly reduces the risk of cross-contamination substitutes deoxyuridine triphosphate (dUTP) for deoxythymidine triphosphate (dTTP) in the amplification procedure. Amplified DNA therefore contains U instead of T and can be degraded by uracil N-glycosylase (UNG). If the PCR reaction solution is first incubated at 50°C in the presence of UNG prior to amplification, the UNG will remove all of the uracil from contaminating previously amplified sequences, rendering them unsuitable as targets for amplification without interfering with the intended target which does not contain uridine. When the reaction mixture is heated to 95°C, the UNG is inactivated so that newly amplified DNA will contain uridine instead of thymidine.

Detection of HIV-1

Most individuals can be diagnosed as being infected with HIV-1 with an enzyme immunoassay (EIA) that detects antibodies to the virus. Repeatedly positive EIAs must be confirmed, usually by Western blot. Detergent disrupted, partially purified HIV-1 is electrophoresed in a polyacrylamide gel to separate the viral proteins. These proteins are electrophoretically transferred to a solid support, such as nitrocellulose, which is cut into strips. These strips serve as the solid phase for an EIA that allows antibody from the patient to react with individual HIV-1 proteins and be visualized. Typically, six to nine characteristic bands are observed if antibodies to HIV-1 are present, and no bands are seen if the antibodies are absent.

Positivity by Western blot requires the presence of antibodies to two of the following HIV-1 proteins: p24, gp41, and either gp120 or gp160 (CDC 1989). A negative result is defined as the absence of reactivity to any bands. Results which fall between these two choices, i.e., some reactive bands but not enough of the specific ones, are deemed indeterminate. When used together, the EIA and Western blot are extremely reliable, with sensitivities of 99.5% and specificities of 99.8% for the diagnosis of HIV-1 infection (Davey and Lane 1990). These tests are fairly rapid, reproducible across laboratories, well quality-controlled and relatively inexpensive, and form the basis of the protection of the blood supply. However, these assays are an indirect measure of infection and are not suitable for all clinical situations.

Detection of HIV-1 in Neonates

Only about 15–30% of the infants born to HIV-1 seropositive mothers are actually infected with the virus, yet all of them will be seropositive at birth due to transplacental transfer of IgG from the mother to the infant. Conventional serological assays cannot discriminate between maternal and infant antibodies, and the maternal IgG can persist in the baby for as long as 18 months. Infected infants often have a much shorter disease course than do adults, and it is extremely important to diagnose infection in these children as early as possible so that treatments can be initiated.

Culture of PBMCs is still considered the "gold standard" for the early diagnosis of HIV-1 infection in neonates. Sensitivity of the assay ranges from 40–50% in newborns, increases to 70–95% by age three months, and after three months is > 95% sensitive (Consensus Workshop 1992). Specificity is close to 100%. However, the HIV culture is time-consuming, labor-intensive, expensive, and requires a BSL 3 laboratory. In addition, at least 2–4 ml of anti-coagulated blood is necessary, and results are not available for 7–28 days.

PCR appears to have similar sensitivity and specificity as culture but has the advantage of being faster (results can often be obtained in 1–2 days). PCR can be performed on as little as 0.5 ml anti-coagulated blood (Casareale et al. 1992) or even on a drop of blood dried on filter paper (Cassol et al. 1992). Cord blood can be used but may be contaminated with maternal blood during collection, resulting in a possible false positive result. Peripheral blood from the child is the preferred source. PCR has not yet been adopted as the "gold standard" because of specificity and quality assurance issues. False positive results due to carry-over contamination which plagued many early experiments are being corrected with modifications in equipment and reagents. The issue of false negative reactions is still being addressed. When these matters are resolved and people become more familiar and confident with the PCR procedure, PCR will probably become the assay of choice.

Detection of HIV-1 in Samples that Give Indeterminate Western Blots

About 5–20% of sera that are repeatedly reactive by EIA are interpreted as indeterminate by Western blot. Indeterminate reactions may be observed early in HIV-1 infection, since antibodies to viral core proteins usually precede the appearance of antibodies to the viral glycoproteins gp41 and gp120/gp160. Other causes of indeterminate Western blots may include loss of anti-core antibodies late in infection, cross-reactive antibodies to HIV-2, cross-reactive antibodies to HLA-DR antigens found in the cells used to propagate the virus that may be found on the Western blot strips, or the presence of autoimmune antibodies.

The CDC has recommended that people who have consistently indeterminate Western blot results for at least six months, in the absence of any known risk factors or clinical findings, should be considered to be negative for antibodies to HIV-1. This recommendation is substantiated by two studies using PCR in patients with indeterminate Western blot results. In the first study, Jackson et al. (1990) found that none of their 99 blood donors had evidence of HIV-1 or HIV-2 infection by PCR or any other assay during the follow-up period which averaged six months, despite the fact that 92% of them continued to have indeterminate Western blot reactions. Celum et al. (1991) studied 89 individuals with indeterminate Western blots. A total of 84 individuals had negative PCR results and continued to be PCR negative during the follow-up period which was greater than six months. Five of the 89 had a positive PCR test. Four of these PCR positive individuals had seroconverted by the next sampling time. The remaining individual who had an initial positive PCR result was negative by PCR when the same specimen was tested by two different laboratories, as were all subsequent specimens from the patient during the next nine months. These two studies demonstrate that current CDC guidelines regarding Western blot are appropriate and PCR is not necessary in these circumstances.

Seronegative Individuals Who Are at Risk for HIV Infection

Several reports have indicated that high-risk individuals (homosexual men, hemophiliacs and their spouses, intravenous drug users) might harbor the virus for several years prior to seroconversion (Ranki et al. 1987; Imagawa et al. 1989; Wolinsky et al. 1989). One study found that 23% of the homosexual men tested who were seronegative were PCR positive (Imagawa et al. 1989). An additional study among the same risk group found individuals who were PCR positive up to 42 months before seroconversion (Wolinsky et al. 1989). Yet more recent studies have refuted these findings, being unable to detect PCR reactivity more than 1–3 months prior to seroconversion (Horsbaugh *et al.* 1990; Lefrere *et al.* 1991; Bruisten *et al.* 1992; Farzedegan *et al.* 1993). Farzedegan *et al.* (1993) recently examined 2159 blood samples from 945 seronegative intravenous drug users. No proviral DNA could be detected in 98.8% of the samples. Specimens from seven individuals were PCR positive, and all five of the patients who were available for follow-up seroconverted within six months.

One explanation for these controversial findings might be false positive PCR reactions (Sheppard *et al.* 1991). Sheppard *et al.* (1991) demonstrated that 8.6% of both their high risk homosexual cohort and their lower risk heterosexual cohort had initially weakly to moderately positive reactions with one or more primer pairs by PCR. All specimens were negative with at least one other primer set, however, and in the final interpretation, 218 of 219 homosexual men and 104 of 105 heterosexual men were deemed to be negative. The remaining two individuals had indeterminate PCR reactivities. In a multicenter study, most of the non-negative PCR results in seronegative individuals were observed with the SK38/39 primer pair, reflecting the widespread use of this set of primers and the increasing potential for contamination of negative samples with previously amplified products (Sheppard *et al.* 1991). Taken together, these results suggest that most individuals seroconvert within a few weeks to months of exposure to the virus and that there is not a lengthy period of time when individuals are PCR positive but seronegative. The three earlier studies reporting such findings may have been flawed with false positive PCR reactions due to carry-over contamination.

HIV-1 Vaccine Recipients

Several HIV-1 vaccines are currently in clinical trials, and more prototypes are being tested in individuals. The antibody response to a killed whole virus vaccine will obscure the detection of those antibodies that might emerge as a result of failure of the vaccine and actual infection. PCR could be used to follow these volunteers to monitor for infection, but this use has not yet been reported.

Application of Quantitative PCR in the Study of the Pathogenesis of HIV-1 Infection

An early study using in situ hybridization suggested that only 1 in 10,000 to 1 in 100,000 PBMC was infected with the virus (Harper *et al.* 1986).

However, several more recent studies using the more sensitive technique of quantitative DNA PCR have demonstrated higher incidences of infected CD4+ cells (Schnittman et al. 1990; Simmonds et al. 1990; Hsai and Spector 1991; Lee et al. 1991; Bagasra et al. 1992; Escaich et al. 1992; Ferre et al. 1992; Jurrians et al. 1992; Connor et al. 1993). In addition, these studies have shown that proviral burden increases with progression of disease. For instance, Schnittman et al. (1990) found that asymptomatic patients had frequencies of HIV-infected CD4 cells of 1/1000 to 1/10,000. However, patients with AIDS, despite similar CD4 cell counts, had frequencies of HIV-infected cells of 1/100 to 1/1000. Similar results have been reported by others (Simmonds et al. 1990; Lee et al. 1991; Escaich et al. 1992), and even higher frequencies have been observed by some investigators (Hsai and Spector 1991; Jurrians et al. 1992). Differences in the frequencies may be due to differences in the methods or to patient differences (such as CD4 cell count or anti-retroviral treatment). In addition, a positive correlation has been observed between the HIV proviral copy number and the infectious HIV titer in PBMC (Escaich et al. 1992; Ferre et al. 1992; Connor et al. 1993).

More recently, in situ PCR has been used to investigate the proportion of PBMCs infected with HIV-1. Adding a PCR step to the in situ hybridization appears to increase the sensitivity of the assay, and two groups of investigators using modifications of this approach now suggest that 1–15% of PBMCs from HIV-1 seropositive individuals carry the HIV genome (Bagasra et al. 1992; Patterson et al. 1993).

Circulating genomic RNA is a reflection of viremia and may be a more accurate measure of events, such as viral replication and viral clearance, which are occurring at given moment in a patient. Most investigators have estimated that the levels of HIV-1 RNA in plasma range from undetectable to 1×10^6 copies/ml (Menzo et al. 1992; Scadden et al. 1992; Bobey et al. 1993; Cao et al. 1993; Dewar et al. 1993; Rasheed et al. 1993). However, Piatek et al. (1993) report values that are generally ten times those for patients at comparable stages of disease. Again, methodologies vary considerably and may be responsible for the differences observed. Regardless of the method used, however, there is general agreement that early in infection, around the time of seroconversion, there seems to be a high level of viremia as measured by p24 antigen, plasma culture and quantitative HIV RNA PCR. As the immune system responds to the infection, viral burden as measured by all assays decreases. During later stages of the disease, as patients become symptomatic, plasma viremia returns and HIV RNA PCR levels increase. The presence of plasma viremia seems to be associated with a more rapidly deteriorating clinical course.

Application of PCR Methods to Clinical Trials of New Anti-retroviral Drugs

Clinical trials of new anti-retroviral drugs involving thousands of patients are being conducted all over the world. The earliest clinical trials included symptomatic patients, many of whom reached a clinical endpoint (a new AIDS defining event or death) during the course of the study. Conclusions about the efficacy of the drug in question could be drawn based on progression of disease and clear clinical data. More recent trials have involved asymptomatic patients who rarely reach a clinical endpoint during the course of the trial. As a consequence, the efficacy of the drugs has been evaluated on surrogate markers of clinical progression and disease. The most widely used and accepted marker has been the CD4 cell count. However, during the past year the correlation between changes in CD4 cell numbers and clinical progression has been questioned. One study observed that changes in CD4 count can only partially explain the clinical benefits of ZDV (Choi et al. 1993). A second study appeared to show that a positive effect of ZDV in CD4 was not reflected in improvement in clinical parameters (Aboulker and Swart 1993). Reasons for these discrepant results and explanations for the lack of correlation between ZDV's effect on CD4 cells and clinical endpoints are being sought. Other immunologic factors might be responsible for the lack of correlation between CD4 cell count and clinical data. An alternative explanation for these findings may be that changes in the viral load, viral genotype or viral phenotype during the long course of infection are more predictive or occur earlier than CD4 changes. Quantitative changes in viral DNA or viral RNA may prove to be better surrogate markers for clinical outcome than CD4, or perhaps the information obtained from these assays may prove complementary to changes in CD4 cell counts (Lefrere et al. 1992).

Quantitative HIV DNA PCR has recently been used to follow patients during anti-retroviral therapy. In one study, treatment with ZDV for 5–14 months did not cause any decrease in the HIV provirus copy number (Donovan et al. 1991). However, two other studies suggested that a marked decrease in HIV proviral DNA occurred following antinucleoside therapy (Aoki et al. 1990; Clark et al. 1992). Aoki et al. (1990) observed a significant decrease in 9 of 12 evaluable patients who received didanosine (ddI) for 8–14 weeks. This decrease was more pronounced in patients receiving higher doses of the drug in this dose escalation Phase I clinical trial. In eight patients treated with ZDV and rIL-2, proviral copy numbers decreased from 100 copies/10^6 PBMC to 13/10^6 PBMC during 20 weeks of treatment (Clark et al. 1992). Proviral copy numbers increased several months after discontinuation of therapy.

All of these studies have involved relatively few numbers of patients (6–13) and employed different methodologies for quantitating the proviral DNA and for performing the PCR. Intra- and interassay variation using positive controls, when reported, ranged from 18–29% and 4–17%, respectively (Clark et al. 1992; Telenti et al. 1992). A considerable amount of variation was observed in untreated, clinically stable individuals during periods ranging from one to sixteen months (Clark et al. 1992). These factors make it extremely difficult to evaluate the use of this assay as a surrogate marker for drug efficacy.

Holodniy et al. (1991) used quantitative RNA PCR to estimate HIV-1 RNA copy number in an investigation of the effect of anti-retroviral drugs in 27 patients. The patients were treated with either ZDV (n = 18), ddI (n = 2), or combination ZDV and ddI (n = 7). After one month of therapy plasma, HIV RNA copy number decreased from 540 ± 175 to 77 ± 35 (p < 0.05). However, the responses of individuals were quite variable. Sixty percent of the patients had a marked decrease in RNA copy number while 11% did not. Similar results were observed in ten patients treated with ddI alone (Aoki-Sei et al. 1992). The reductions in HIV-1 RNA were considerably more substantial and significant than the decreases observed in HIV-1 DNA in the same patients (Aoki et al. 1990).

Using a competitive RNA PCR, Piatek et al. (1993) found that treatment with ZDV for 1–20 weeks resulted in up to 39-fold decreases in the amount of HIV-1 RNA. Three previously untreated patients were given ZDV for six weeks, followed by one week off therapy. Rapid decreases were seen in all three patients after only one week of treatment, but levels of circulating HIV-1 RNA rebounded to pre-treatment levels within one week after discontinuation of the drug. The average decrease in HIV-1 RNA among the ZDV treated patients was 11-fold.

Molecular Techniques in HIV
Drug-susceptibility Testing and Resistance

The rapid discovery that a human retrovirus was the etiologic agent of AIDS led to the equally rapid development of compounds designed to inhibit replication of the virus. Many of these compounds have shown activity in vitro and some, in particular the nucleoside analogue reverse transcriptase (RT) inhibitors zidovudine (ZDV), didanosine (ddI) and zalcitibine (ddC), have demonstrated antiretroviral activity and limited clinical efficacy in humans infected with HIV. Non-nucleoside RT inhibitors (NNRTI) have also shown antiretroviral activity in HIV-1 infected persons. However, HIV-isolates resistant to ZDV, other nucleoside RT inhibitors, NNRTI's and other classes of anti-HIV agents have been isolated in vitro and in vivo. The emergence of ZDV-resistant isolates has been correlated with disease progression, but a causal relationship has not been proven (Tudor-Williams et al. 1992; Kozal et al. 1993; AIDS Clinical Trial Group 1993). Techniques of molecular biology have been used extensively in the research laboratory to help elucidate the genetic mechanisms of HIV-1 resistance to ZDV and other reverse transcriptase inhibitors (Larder et al. 1989; Larder and Kemp 1989; Larder et al. 1991; Nunberg et al. 1991; Richman et al. 1991; St. Clair et al. 1991; Boucher et al. 1992; Fitzgibbon et al. 1992; Gu et al. 1992; Richman and the ACTG 164/168 Study Team 1992; Vasudevachari et al. 1992; Chow et al. 1993; Eron et al. 1993; Mellors et al. 1993; Schinazi et al. 1993). These methods have included but are not limited to PCR amplification and sequence analysis (Larder et al. 1989; St. Clair et al. 1991; Kellam et al. 1992; Eron et al. 1993), site specific mutagenesis and molecular cloning (Larder and Kemp 1989; St. Clair et al. 1991; Fitzgibbon et al. 1992; Kellam et al. 1992; Eron et al. 1993), in situ hybridization (Richman et al. 1991), and mutation-specific PCR (Larder et al. 1991; Boucher et al. 1992). Direct sequencing of PCR products (Casanova et al. 1990) has facilitated the detection of HIV-1 drug-resistance mutations (Jung et al. 1992; Eron et al. 1993).

Drug Susceptibility Testing

Developing a test that can document emergence of resistant virus in patients infected with HIV on a scale and at a cost useful to a clinical virology laboratory remains a challenge. The methods of conventional virology currently used to determine the drug-susceptibility phenotype of

HIV isolates are time-consuming and expensive. A recently published consensus (Japour *et al.* 1993) that outlines a very efficient cell culture assay for detecting virus susceptibility phenotype requires at least 14 days to give results and costs approximately $300–500 per assay (personal communication S. Fiscus). In addition, most techniques require some growth of the viral isolate in the laboratory, which may result in a change in the predominant HIV-1 genotype. Investigators have used molecular techniques to try to improve and shorten the process for determining HIV-1 drug-susceptibility phenotype.

Eron *et al.* (1992) used quantitative PCR to detect HIV-1 DNA to measure replication of clinical isolates of HIV in vitro in the presence of nucleoside RT inhibitors. The PCR product was detected with the previously described enzyme-linked oligonucleotide sandwich assay (ELOSA, Du Pont, Billerica, MA), and quantitation was accomplished by comparison to a standard curve generated from a chronically infected cell line known to contain one copy of HIV-1 proviral DNA per cell. The results of the assay could be obtained within 48–72 hours, and correlation with conventional p24 antigen based tissue culture was good (Eron *et al.* 1993). Winters *et al.* (1992) have also used gene amplification of both HIV-1 DNA and RNA to measure replication of HIV-1 clinical isolates in vitro to assess drug-susceptibility. In both assay systems, cultures could be inoculated with virus at a low multiplicity of infection, thereby limiting the number of in vitro passages of virus needed prior to susceptibility testing. However, both assays require a large number of amplifications, and neither assay has gained wide use beyond the institution at which it was developed.

Mutation-specific PCR

The molecular basis of the resistant phenotype is known for many of the agents to which HIV-1 has developed resistance. Molecular techniques that can rapidly and specifically detect these mutations in clinical samples may help clinical investigators and clinicians quickly determine whether a patient has developed drug-resistant HIV-1. Five separate reverse transcriptase gene mutations have been shown to be associated with ZDV-resistance (Larder *et al.* 1989; Larder and Kemp 1989; Kellam *et al.* 1992). However, two separate mutations at codon 215 in the RT region of the HIV-1 *pol* gene, each a two base-pair change resulting in one of two single amino acid substitutions (Thr to Tyr or Phe), have been shown to be the

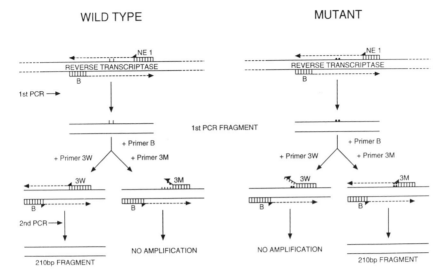

Figure 11.4. Mutation-specific PCR to detect ZDV-resistance mutation at codon 215. A primer pair (B, NE1) designed to anneal to conserved sequences that flank the area on the RT region of the *pol* gene containing the 215 mutation is used in initial PCR reaction. The second PCR step uses product from the first reaction mixture. Two separate reactions using the first PCR fragment are then performed using one of the conserved primers (primer B) used in the first reaction, paired with one of two primers (3W or 3M) that is designed to anneal with its 3' end overlying the 215 coding region. The reaction with the primer designed to anneal to the wild-type sequence at codon 215 (3W) will yield a 210 bp product if the wild-type coding sequence is present in the reaction mixture. The second reaction with the primer that would anneal to mutant sequence at codon 215 will yield product if either of the two mutant sequences are present. Products of the two reaction were separated on a agarose gel and identified by ethidium bromide staining. Similar methods with second reaction primers directed to wild-type and mutant sequence of other ZDV-resistance loci and loci for other drug-resistance mutations have been described.

most common ZDV resistance mutation. A mutation at codon 215 is present almost uniformly when a high level of resistance occurs (Richman *et al.* 1991; Boucher *et al.* 1992).

Mutation-specific PCR was first used to study HIV-1 resistance to ZDV by a collaborative group of Dutch and British investigators (Boucher *et al.* 1990). Using a "nested" PCR reaction (Figure 11.4), the investigators first used a primer pair (B, NE1) designed to anneal to conserved sequences that flanked the area on the RT gene containing either 215 mutation. In the second PCR step a 210 bp fragment was amplified using one of the conserved primers used in the first reaction, paired with one of two primers (3W or 3M) that was designed to anneal to the RT gene with its 3' end overlying the 215 coding region. Two parallel reactions were performed: one with the primer designed to anneal to the wild-type sequence at codon 215 (3W), and the second with the primer that would anneal to mutant sequence at codon 215. Products of the two reactions were separated on an agarose gel and identified by ethidium bromide staining. Samples for study were obtained from patients treated with ZDV by culturing patients' PBMC with PHA-stimulated PBMC from HIV-seronegative donors and then extracting DNA from these "cocultured" samples. Samples were scored as containing wild-type HIV-1 when the 210 bp fragment obtained by PCR product amplified with the wild-type primer (3W) stained most intensely with ethidium bromide. The sample was considered to contain mutant forms when the most intensely staining band was observed after amplification with mutant primer (3M). When bands from both of the second PCR reactions were of equal intensity, the sample was judged to contain a mixture of wild-type and mutant viruses.

The specificity of amplification with this nested primer would depend in part on the stringency of annealing. Specificity by this criterion appeared adequate for wild-type sequence, as all isolates obtained prior to ZDV therapy contained wild-type HIV-1. Sixteen out of 18 individuals had mutant virus after 48–112 weeks of therapy; however, no confirmatory sequencing data were presented to determine exact specificity and sensitivity for detection of mutant sequence at codon 215. The drug-susceptibility phenotype was established in only a minority of patients, though correlation of phenotypic resistance with the presence of the 215 mutation was good. Four out of 16 isolates obtained at an intermediate time point during therapy had signals of equal intensity after amplification with both the wild-type and mutant primer pairs. The authors postulated that these isolates contained mixtures of wild-type and mutant virus. They also pointed out, however, that the coding sequence of an intermediate amino acid, such as serine at codon 215, might anneal successfully with both the wild-type and mutant primers and give bands of equal intensity in each of the second PCR reactions.

The relative contribution of each of the mutations to ZDV resistance is important to determine, especially if degree of ZDV sensitivity is going to be assessed by mutation-specific PCR alone. In 1991 Larder et al.

described the use of mutation-specific PCR in detecting each of the four ZDV resistance mutations known at that time. Primer pairs that amplify regions containing the wild-type or the mutant sequence at each of the four ZDV resistance alleles were created. Nested PCR was required to use mutation-specific PCR directly on DNA from PBMC from HIV-infected individuals. As outlined previously, primers that annealed to the mutant and wild-type sequence were used in parallel reactions for each of the RT gene codons evaluated. For each set of primers used to detect the wild-type or mutant sequence at a specific allele, conditions had to be such that specific priming was enhanced and nonspecific priming limited. These conditions were achieved by varying $MgCl^2$ concentrations, varying annealing temperatures and, for some of the primers, deliberately mismatching the sub 3' terminal base-pair to enhance selectivity. The addition of 1 µg of herring DNA to each reaction also improved selectivity. Notably, sequence for either mutation at codon 215 (Phe or Tyr) could be detected using a single primer that had complimentary sequence(3'AAG) to the Phe codon sequence (5'TTC). Using patient samples, the results of mutation-specific PCR for codons 70 and 215 were compared with both sequence analysis and the results of ZDV susceptibility testing. Using site-specific mutagenesis, codon 215 was shown to contribute significantly to the degree of ZDV resistance, though the range of susceptibility of clones that contained the 215 mutation was wide and depended on the number and location of additional mutations. Importantly, the authors were able to show that the ZDV-resistance mutations at codons 70 and 215 could be detected directly in PBMC from infected individuals without a co-culture step. They also pointed out that this technique can detect low levels of resistance mutations even when the wild-type is in a thousand-fold excess. The clinical utility of detecting such a minority species is unclear.

Richman et al. (1991) used PCR and radio-labelled oligonucleotide probes to detect ZDV resistance mutations. Using PBMC from infected patients, or virus isolated from subjects by in vitro co-cultivation, these investigators amplified a 535 bp fragment using primers flanking conserved areas on the RT gene. Amplified DNA was then separated by size in agarose gels, transferred to a nylon membrane, and probed with one of eight radio-labelled oligonucleotide probes, a wild-type and mutant probe for each of the four known ZDV-resistance codons. This assay was shown to be extremely sensitive, able to detect wild-type or mutant alleles in diluted samples containing one infected cell or one tissue culture infectious dose (TCID50) in a cell-free solution. The PCR/oligonucleotide probe assay could detect signal in all cell free virus stocks (using a reverse transcription step) and 99% of all the PBMC samples from infected subjects. All pre-ZDV therapy samples reacted with all four wild-type probes. All samples obtained after ZDV therapy hybridized to at least one of the eight probes, confirming successful amplification of HIV DNA. However, neither the wild type nor the mutant probe hybridized at one or more of the four loci in 55 of 237 samples, suggesting that mutations other than those for which the probes were designed may have occurred in or around the codon of interest, hence interfering with probe hybridization. Detection of virus of known genotype was specific. Using this method and concurrent drug-susceptibility testing, the authors examined 304 independent specimens from 168 individuals treated with ZDV to evaluate ZDV resistance development over time. Mutations at codons 70 and 215 were noted to appear first and to be the most common mutations after one year of ZDV therapy. Mixtures of mutant and wild-type sequence from subjects on ZDV therapy were also documented. The degree of drug-susceptibility correlated with the number of ZDV resistance mutations present, though a range of ZDV susceptibilities were documented by phenotypic testing, even for a given single mutation or a given set of mutations. The presence of a range of ZDV susceptibilities when their assay documented a specific combination of wild-type and mutant alleles at the four loci being tested led the authors to postulate that additional ZDV resistance mutations existed. Subsequent to the study of Richman et al. (1991), a fifth ZDV resistance mutation was discovered (Kellam et al. 1992). The assay proved useful for studying general patterns of ZDV resistance. However, the failure of wild-type or mutant probe to hybridize at one or more codons in 23% of the samples obtained on therapy and the inability to detect as yet undescribed mutations limit the utility of this assay to studying individual patients in the clinical laboratory.

Boucher et al. (1992) used the mutation-specific assay outlined earlier (Larder et al. 1991) to examine the natural history of ZDV resistance in 18 patients treated with ZDV. Mutations at codons 70 and 215 appeared first in the patients studied. Once the 215 mutation appeared, the mutation at codon 70 could revert to wild-type. An association between the number of mutations detected by mutation-specific PCR and the degree of resistance documented in a phenotypic assay was noted in three patients who were studied with both methods. No clear correlation between development of intermediate resistance as assessed by the mutation-specific assay and disease progression was noted. High-level resistance occurred in only one patient after the development of AIDS.

Mutation-specific PCR primarily focusing on the 215 muation is being used to evaluate larger numbers of patient samples to try to correlate the appearance of this ZDV resistance mutation with clinical progression. Kozal et al., (1993) using the methods of Larder et al. (1991), obtained DNA from cryopreserved PBMC and RNA from stored serum from 38 minimally symptomatic patients treated for a mean of 34 months with ZDV. The authors processed all samples in duplicate and included positive, negative (wild-type and mutant), and reaction mixture controls with each run. Non-selective priming was seen if input HIV DNA was of high copy number, and this possibility was controlled for. Of 38 patients, 45% had virus mutant at codon 215 detected in DNA from PBMC, while 55% had only wild-type detected. There was no correlation between length of therapy or starting CD4 count with the appearance of the 215 mutation. However, at the end of the study, the mean CD4 of patients with the 215 mutation was more than 50% lower than those who were wild-type ($p < 0.0001$). The data obtained from the RNA samples are more dificult to interpret as the reverse transcription step performed prior to the nested PCR may have been of low specificity. Overall, the authors showed an association between the appearance of the 215 mutation and a decline in CD4 counts. Contrary to earlier studies (Boucher et al. 1990; Richman et al. 1991), over half of the patients' therapy remained wild-type at codon 215 despite a mean duration of ZDV therapy of 34 months. Other groups are using codon 215 mutation-specific PCR on even larger numbers of patients treated with more complex antiretroviral regimens (Larder and the Protocol 34 225-02 Collaborative Group 1993; Shafer et al. 1993).

One significant issue concerning mutation-specific PCR is the issue of quality assurance. Sensitivity and specificity of the assay for detecting the specific mutation must be measured, but also the specificity and sensitivity of the presence of the mutation in predicting a resistant phenotype or clinical progression must be known in order for the assay to have utility in the clinical laboratory. The AIDS Clinical Trials Group (ACTG) Virology Committee has developed a protocol for the detection of the 215 mutation (personal comunication, Robert W. Shafer) based on the technique of Larder and Boucher (1993). Using specimens from a large clinical trial (Kahn *et al.* 1992), co-cultured PBMC were assayed for the presence of the 215 mutation, and these results were correlated with the phenotype as determined by the ACTG consensus protocol (Japour *et al.* 1993). The sensitivity of the 215 PCR assay for detecting high level ZDV resistance, defined as an $IC^{50} \geq 0.1$ μM, was 85%, but the specificity was only 59%. Twelve out of 29 isolates with $IC^{50} < 0.1$ μM were detected as mutant, including five with IC^{50}'s of 0.001 μM, a phenotype which would be consider completely sensitive to ZDV (Kuritzkes *et al.* 1993). Sequence data suggest that the specificity of the assay for the 215 mutation itself was much higher, implying that the presence of the 215 mutation does not always correlate with phenotypic ZDV resistance (personal communication, Daniel R. Kuritzkes). This finding may be due to the interactive effects of other nucleoside resistant mutations at codon 215 (St. Clair *et al.* 1991; Eron *et al.* 1993).

Overall, molecular techniques, such as mutation-specific PCR or specific oligonucleotide hybridization to PCR amplified products, can only be useful in detecting known mutations. The wide range of phenotypic resistance present when a specific resistance mutation is documented may limit the usefulness of these assays in predicting the degree of resistance in an individual patient. This range may be due to undetected mutations at other coding sites that affect the phenotype observed when the sought-after mutation is detected. Resistance of HIV to even one agent may be conferred by one or a combination of mutations or may be lessened or enhanced by a mutation at another site, selected for by a different agent, as reviewed in Richman 1993. The usefulness of mutation-specific assays may be limited to the study of different populations of patients treated with various ZDV or combination therapies. These assays could document over time the development of ZDV resistance mutations and implicate or refute the causal association of ZDV resistance mutations with clinical progression. These mutation-specific assays could also be used to stratify patients prior to entry into clinical trials. As additional agents are developed and patients are treated with combinations of medications, the number of mutations associated with single or multiple drug-resistance will continue to increase. Preliminary data are being gathered to assess whether mutation-specific PCR can detect resistance mutations to agents other than ZDV (Campbell *et al.* 1993; Larder and Boucher 1993). As the number of clinically useful drug agents increases and more resistance mutations to these agents are described, the utility of mutation-specific PCR may diminish because of the number of loci that may need to be investigated in subjects on multiple drug combination therapy.

Validation and Quality Assurance of Quantitative PCR Techniques

Quality control and quality assurance are mandatory in a clinical laboratory. Several positive and negative controls should be scattered randomly within each run. Adequate sensitivity is indicated by the consistent detection of a 5–10 copy number assay standard and the positive controls. Adequate specificity is indicated by the lack of false positives in the negative controls. Primer pairs specific for a single copy cellular gene should be included as an internal control for the amplification process. Commonly used as internal controls are the HLA DQ-alpha or beta-globin genes. Absence of a specific band for the internal control indicates that the PCR has not worked and must be repeated for that sample. Several national and international groups have begun standardization and proficiency testing programs for PCR (Sheppard *et al.* 1991; Bootman and Kitchin 1992; Defer *et al.* 1992). Even with the use of standardized reagents, sensitivities and specificities among some of the participating laboratories varied considerably. More consistent results with fewer false positives must be obtained before this methodology becomes widely accepted in the clinical laboratory.

Laboratories must establish criteria for interpreting the results of a specific PCR assay so that results can be compared across laboratories. The following algorithm has been proposed for detection of HIV-1 sequences in pediatric patients (Consensus Workshop 1992). Blood specimens should be tested in duplicate, preferably with more than one primer pair. Any sample with discordant or indeterminate results should be retested to resolve the discrepancies before results are reported. For a positive PCR result, a second specimen should be obtained to confirm the diagnosis. The AIDS Clinical Trials Group of the NIH currently recommends that diagnosis of perinatal infection by PCR be confirmed by an additional test (p24 antigen or virus isolation) on a second specimen (AIDS Clinical Trials Group 1993). This recommendation may change as techniques to avoid cross-contamination and carry-over are improved.

Almost all descriptions of quantitative PCR methodologies address the sensitivity and specificity of the particular technique. However, validation of quantitative PCR methods should also include the evaluation of the accuracy, precision, reproducibility and linear range of an assay. It is important to remember that the relevant sensitivity limit of a quantitative assay is the limit of quantitation rather than the limit of detection, that is, the capability of measuring precisely and/or accurately the number of HIV DNA or RNA copies versus the ability of detecting the lower limits of the assay. Too little attention to date has been placed on the amount of variation in the quantitative PCR methodologies. In at least two studies, the PCR was found not to be reproducible enough or sufficiently precise to allow discrimination of differences in HIV copy numbers (Davis et al. 1990; Oka et al. 1992).

For the purposes of monitoring clinical trials, it may only be necessary to estimate and compare relative amounts of HIV nucleic acid instead of absolute copy numbers. If this is the case, it is more important to determine an assay's variability instead of accuracy. Variability is a function of precision (the amount of interassay variation) and reproducibility (the amount of intra-assay variation). Intra- and interassay variability for quantitative PCR has been reported to be 5–30% (Davis et al. 1990; Dickover et al. 1990; Clark et al. 1992; Oka et al. 1992; Telenti et al. 1992).

As simpler and better controlled techniques are devised, as the procedure becomes more automated, and as technologists gain more experience with the test, measurement variation should decrease and make this assay more suitable for monitoring drug efficacy and progression of disease. Quality assurance and proficiency testing will play an important role in establishing quantitative PCR assays in multi-center clinical trials. Part of the quality assurance program will have to address the inter- and intra-assay and inter-laboratory variation and establish statistical limits of significance.

Summary

The use of molecular techniques in the clinical antiretroviral laboratory shows promise in several areas. Detection by PCR of HIV-1 infection prior to seroconversion or in infants born to HIV-infected mothers is approaching wide clinical application. Quantification of viral load as reflected by plasma RNA levels measured with RNA PCR or branch DNA techniques are actively being evaluated for clinical utility. The use of specific molecular techniques to identify resistance development to antiretroviral agents requires further investigation prior to their use in the clinical laboratory.

References

Aboulker, J.-P. and Swart, A. 1993. "Preliminary analysis of the Concorde trial." *Lancet* 341: 889-90.

AIDS Clinical Trial Group, June 1, 1993. "ACTG 116B117: Preliminary resistance analysis, Summary for study investigators." Bethesda, MD,

AIDS Clinical Trials Group, 1993. "Virology Manual." Bethesda, MD,

Aoki, S., Yarchoan, R., Thomas, R.V., Pluda, J.M., Marczyk, K., Broder, S. 1990. "Quantitative analysis of HIV-1 proviral DNA in peripheral blood mononuclear cells from patients with AIDS or ARC: decrease of proviral DNA content following treatment with 2',3'-dideoxyinosine (ddI)." *AIDS Res Human Retroviruses* 6: 1331-9.

Aoki-Sei, S., Yarchoan, R., Kageyama, S., *et al.* 1992. "Plasma HIV-1 viremia in HIV-1 infected individuals assessed by polymerase chain reaction." *AIDS Res Human Retroviruses* 8: 1263-70.

Bagasra, O., Hauptman, S., Lischner, H., Sachs, M., Pomerantz, R. 1992. "Detection of human immunodeficiency virus type 1 provirus in mononuclear cells by in situ polymerase chain reaction." *N Engl J Med* 326: 1385-91.

Barre-Sinoussi, F., Chermann, J., Rey, F., *et al.* 1983. "Isolation of T-lymphotrophic retrovirus from a patient at risk for acquired immune deficiency syndrome (AIDS)." *Science* 220: 868-71.

Bobey, L., Saxer, M., Kokka, R., Feinberg, M., Volberding, P., Elbeik, T. 1993. "Quantitation of plasma derived HIV-1 virion RNA using the HIV-1 branched DNA (bDNA) assay." *IX International Conference on AIDS*, Berlin, Abstract PO-B41-2485.

Bootman, J., Kitchin, P. 1992. "An international collaborative study to assess a set of reference reagents for HIV-1 PCR." *J Virol Methods* 37: 23-42.

Boucher, C. A., *et al.* 1992. "Ordered appearance of zidovudine resistance mutations during treatment of 18 human immunodeficiency virus-positive subjects." *J Infect Dis* 165(1): 105-10.

Boucher, C. A., *et al.* 1990. "Zidovudine sensitivity of human immunodeficiency viruses from high-risk, symptom-free individuals during therapy." *Lancet* 336(8715): 585-90.

Bruisten, S., Koppelman, M., Dekker, J., *et al.* 1992. "Concordance of human immunodeficiency virus detection by polymerase chain reaction and by serologic assays in a Dutch cohort of seronegative homosexual men." *J Infect Dis* 166: 620-2.

Campbell, T. B., Routh, J. A., Bakhtiari, M., Schooley, R. T., Kuritzkes, D. R., June, 1993. "Detection of Genotypic Changes in Reverse Transcriptase during combination therapy with zidovudine and L-697-661." *IX International Conference on AIDS*, Berlin, PO-A26-0630.

Candotti, D., Jung, M., Kerouedan, D., *et al.* 1991. "Genetic variability affects the detection of HIV by polymerase chain reaction." *AIDS* 5: 1003-7.

Cao, Y., Kokka, R., Kern, D., Urdea, M., Wu, Y., Ho, D. 1993. "Comparison of quantitative bDNA technique with end-point dilution culture, p24 antigen assay, and RT-PCR quantitation of HIV-1 in plasma." *IX International Conference on AIDS*, Berlin, Abstract PO-A32-0794.

Casanova, J.-L., Pannetier, C., Jaulin, C., Kourilsky, P. 1990. "Optimal conditions for directly sequencing double-stranded PCR products with sequenase." *Nucl Acids Res* 18: 4028.

Casareale, D., Pottathil, R., Diaco, R. 1992. "Improved blood sample processing for PCR." *PCR Methods Application* 2: 149-53.

Cassol, S., Lapointe, N., Salas, T., *et al.* 1992. "Diagnosis of vertical HIV-1 transmission using the polymerase chain reaction and dried blood spot specimens." *J AIDS* 5: 113-9.

CDC 1989. "Interpretation and use of the Western blot assay of serodiagnosis for human immunodeficiency virus type 1 infections." *MMWR* 38(S7): 1-7.

Celum, C., Coombs, R., Lafferty, W., *et al.* 1991. "Indeterminate HIV-1 Western blot: Seroconversion risk, specificity of supplemental tests, and an algorithm for evaluation." *J Infect Dis* 164: 656-64.

Choi, S., Lagakos, S., Schooley, R., Volberding, P. 1993. "CD4+ lymphocytes are an incomplete surrogate marker for clinical progression in persons with asymptomatic HIV infection taking zidovudine." *Ann Intern Med* 118: 674-80.

Chomczynski, P., Sacchi, N. 1987. "Single-step method of RNA isolation by acid guanidinium thiocyanate-phenol-chloroform extraction." *Anal Biochem* 162: 156-9.

Chow, Y. K., *et al.* 1993. "Use of evolutionary limitations of HIV-1 multidrug resistance to optimize therapy." *Nature* 361(6413): 650-4.

Clark, A., Holodniy, M., Schwartz, D., Katzenstein, D., Merigan, T. 1992. "Decrease in HIV provirus in peripheral blood mononuclear cells during zidovudine and human rIL-2 administration." *J AIDS* 5: 52-9.

Clewley, J. 1989. "The polymerase chain reaction, a review of the practical limitations for human immunodeficiency virus diagnosis." *J Virol Methods* 25: 179-88.

Connor, R., Mohri, H., Cao, Y., Ho, D. 1993. "Increased viral burden and cytopathicity correlate temporally with CD4+ T-lymphocyte decline and clinical progression in human immunodeficiency virus type 1-infected individuals." *J Virol* 67: 1772-7.

Consensus Workshop 1992. "Early diagnosis of HIV infection in infants." *J AIDS* 5: 1169-78.

Coutlee, F., Viscidi, R., Saint-Antoine, P., Kessous, A., Yolken, R. 1991. "The polymerase chain reaction: a new tool for the understanding and diagnosis of HIV-1 infection at the molecular level." *Mol Cell Probes* 5: 241-59.

Davey, R., Lane, C. 1990. "Laboratory methods in the diagnosis and prognostic staging of infection with human immunodeficiency virus type 1." *Rev Infect Dis* 12: 912-30.

Davis, G., Blumeyer, K., DiMichele, L., *et al.* 1990. "Detection of human immunodeficiency virus type 1 in AIDS patients using amplification-mediated hybridization analyses: reproducibility and quantitation limitations." *J Infect Dis* 162: 13-20.

Defer, C., Agut, H., Garbarg-Chenon, A., *et al.* 1992. "Multicenter quality control of polymerase chain reaction for detection of HIV DNA." *AIDS* 6: 659-63.

Dewar, R., *et al.*, June, 1993. "Comparison of branched DNA (bDNA) technology with virus culture for quantitation of HIV-1 in plasma." *IX International Conference of AIDS*, Berlin, Abstract PO-B41-2476.

Dickover, R., Donovan, R., Goldstein, E., Dandekar, S., Bush, C., Carlson, J. 1990. "Quantitation of Human Immunodeficiency Virus DNA by using the polymerase chain reaction." *J Clin Microbiol* 28: 2130-2133.

Donovan, R., Dickover, R., Goldstein, E., Huth, R., Carlson, J. 1991. "HIV-1 proviral copy number in blood mononuclear cells from AIDS patients on zidovudine therapy." *J AIDS* 4: 766-9.

Eron, J., *et al.* 1993. "*pol* Mutations conferring Zidovudine and didanosine resistance with different effects in vitro multiply resistant human immunodeficiency virus type 1 isolates in vivo." *Antimicrob Agents Chemother* 37(7): 1480-7.

Eron, J. J., Gorczyca, P., Kaplan, J. C., D'Aquila, R. T. 1992. "Susceptbility testing by polymerase chain reaction DNA quantitation: A method to measure drug resistance of human immunodeficiency virus type 1 isolates." *Proc Natl Acad Sci USA* 89: 3241-3245.

Escaich, S., Ritter, J., Rougier, P., *et al.* 1992. "Relevance of the quantitative detection of HIV proviral sequences in PBMC of infected individuals." *AIDS Res Human Retroviruses* 8: 1833-7.

Farzedegan, H., Vlahov, D., Solomon, L., *et al.* 1993. "Detection of human immunodeficiency virus type 1 infection by polymerase chain reaction in a cohort of seronegative intravenous drug users." *J Infect Dis* 168: 327-31.

Ferre, F. 1992. "Quantitative or semi-quantitative PCR: Reality versus myth." *PCR Methods Applications* 2: 1-9.

Ferre, F., Marchese, A., Duffy, P., *et al.* 1992. "Quantitation of HIV viral burden by PCR in HIV seropositive Navy personnel representing Walter Reed stages 1 to 6." *AIDS Res Human Retroviruses* 8: 269-75.

Fitzgibbon, J. E., Howell, R. M., Haberzettl, C. A., Sperber, S. J., Gocke, D. J., Dubin, D. T. 1992. "Human immunodeficiency virus type 1 pol gene mutations which cause decreased susceptibility to 2',3'-dideoxycytidine." *Antimicrob Agents Chemother* 36(1): 153-7.

Gu, Z., Gao, Q., Li, X., Parniak, M. A., Wainberg, M. A. 1992. "Novel mutation in the human immunodeficiency virus type 1 reverse transcriptase gene that encodes cross-resistance to 2',3'-dideoxyinosine and 2',3'-dideoxycytidine." *J Virol* 66(12): 7128-35.

Harper, M. E., Marselle, L. M., Gallo, R. C., Wong-Staal, F. 1986. "Detection of lymphocytes expressing human T-lymphotropic virus type III in lymph nodes and peripheral blood from infected individuals by *in situ* hybridization." *Proc Natl Acad Sci USA* 83: 772-776.

Henrard, D., Mehaffey, W., Allain, J.-P. 1992. "A sensitive viral capture assay for detection of plasma viremia in HIV-infected individuals." *AIDS Res Human Retroviruses* 8: 47-52.

Holodniy, M., Katzenstein, D., Israelski, D., Merigan, T. 1991. "Reduction in plasma human immunodeficiency virus ribonucleic acid after dideoxynucleoside therapy as determined by the polymerase chain reaction." *J Clin Invest* 88: 1755-9.

Holodniy, M., Kim, S., Katzenstein, D., Konrad, M., Groves, E., Merigan, T. 1991. "Inhibition of human immunodeficiency virus gene amplification by heparin." *J Clin Microbiol* 29: 676-9.

Horsbaugh, C., Ou, C., Jason, J., *et al.* 1990. "Concordance of polymerase chain reaction with HIV antibody detection." *J Infect Dis* 162: 542-5.

Hsai, K., Spector, S. 1991. "Human immunodeficiency virus DNA is present in a high percentage of CD4+ lymphocytes of seropositive individuals." *J Infect Dis* 164: 470-5.

Imagawa, D., Lee, M., Wolinsky, S., *et al.* 1989. "Long latency of human immunodeficiency virus-1 in seronegative high risk homosexual men determined by prospective virus isolation and DNA amplification studies." *N Engl J Med* 320: 1458-62.

Jackson, J. 1993. "Detection and quantitation of human immunodeficiency virus type 1 using molecular DNA/RNA technology." *Arch Pathol Lab Med* 117: 473-7.

Jackson, J., MacDonald, K., Cadwell, J., *et al.* 1990. "Absence of HIV infection in blood donors with indeterminate Western blot test for antibody to HIV-1." *N Engl J Med* 322: 217-22.

Jackson, J., Ndugwa, C., Mmiro, F., *et al.* 1991. "Non-isotopic polymerase chain reaction methods for the detection of HIV-1 in Ugandan mothers and infants." *AIDS* 5: 1463-7.

Japour, A., et al. 1993. "Standardized peripheral blood monocuclear cell culture assay for determination of drug susceptibilities of clinical human immunodefiency virus type 1 isolates." Antimicrob Agents Chemother 37(5): 1095-1101.

Jung, M., Agut, H., Candotti, D., Ingrand, D., Katlama, C., Huraux, J. M. 1992. "Susceptibility of HIV-1 isolates to zidovudine: correlation between widely applicable culture test and PCR analysis." J Acquir Immune Defic Syndr 5(4): 359-64.

Jurrians, S., Dekker, J., deRonde, A. 1992. "HIV-1 viral DNA load in peripheral blood mononuclear cells from seroconverters and long-term infected individuals." AIDS 6: 635-41.

Kahn, J. O., et al. 1992. "A controlled trial comparing continued zidovudine with didanosine in human immunodeficiency virus infection." N Engl J Med 327: 581-587.

Katzenstein, D.A., Holodniy, M., Israelski, D.M., et al. 1992. "Plasma viremia in human immunodeficiency virus infection: relationship to stage of disease and antiviral treatment." J AIDS 5: 107-12.

Kellam, P., Boucher, C. A., Larder, B. A. 1992. "Fifth mutation in human immunodeficiency virus type 1 reverse transcriptase contributes to the development of high-level resistance to zidovudine." Proc Natl Acad Sci USA 89(5): 1934-8.

Kozal, M. J., Shafer, R. W., Winters, M. A., Katzenstein, D. A., Merigan, T. C. 1993. "A mutation in human immunodeficiency virus reverse transcriptase and decline in CD4 lymphocyte numbers in long-term zidovudine recipients." J Infect Dis 167(3): 526-32.

Kuritzkes, D., et al., June, 1993. "Baseline prevalence and clinical significance of zidovudine (ZDV) resistance mutations in HIV-1 isolates from patients participating in ACTG protocol 116B/117." HIV Drug Resistance—Second International Workshop, Noordwijk, The Netherlands.

Kwok, S., Higuchi, R. 1989. "Avoiding false positives with PCR." Nature 339: 237-8.

Kwok, S., Mack, D., Mullis, K., et al. 1987. "Identification of human immunodeficiency virus sequences by using in vitro enzymatic amplification and oligomer cleavage detection." J Virol 61: 1690-4.

Larder, B. A., Boucher, C. A. B. (1993). PCR detection of Human Immunodeficiency Virus drug resistance mutations. *Diagnostic Molecular Microbiology*. In Press.

Larder, B. A., Darby, G., Richman, D. D. 1989. "HIV with reduced sensitivity to zidovudine (AZT) isolated during prolonged therapy." *Science* 243: 1731-1734.

Larder, B. A., Kellam, P., Kemp, S. D. 1991. "Zidovudine resistance predicted by direct detection of mutations in DNA from HIV-infected lymphocytes." *AIDS* 5(2): 137-44.

Larder, B. A., Kemp, S. D. 1989. "Multiple mutations in the HIV-1 reverse transcriptase confer high-level resistance to zidovudine (AZT)." *Science* 246: 1155-1158.

Larder, B. A. and the Protocol 34 225-02 Collaborative Group, June, 1993. "Virological analysis of HIV isolates from individuals enrolled in a placebo-controlled trial of zidovudine alone, or in combination with ddI or ddC." *IX International Conference on AIDS*, Berlin, Abstract WS-B25-2.

Lee, T.-H., Sunzeri, F., Tobler, L., WIlliams, B., Busch, M. 1991. "Quantitative assessment of HIV-1 DNA load by coamplification of HIV-1 gag and HLA-DQ-alpha genes." *AIDS* 5: 683-91.

Lefrere, J., Mariotti, M., Vittecoq, D. 1991. "No evidence of frequent HIV-1 infection in seronegative at-risk individuals." *Transfusion* 31: 205-11.

Lefrere, J.-J., Mariotti, M., Wattel, E., et al. 1992. "Towards a new predictor of AIDS progression through the quantitation of HIV-1 DNA copies by PCR in HIV-infected individuals." *Br J Haematol* 1992: 467-71.

Loche, M., Mach, B. 1988. "Identification of HIV-infected seronegative individuals by a direct diagnostic test based on hybridization to amplified viral DNA." *Lancet* 418(ii): 418-21.

Mellors, J. W., et al. 1993. "A single conservative amino acid substitution in the reverse transcriptase of human immunodeficiency virus-1 confers resistance to (+)-(5S)-4,5,6,7-tetrahydro-5-methyl-6-(3-methyl-2-butenyl)imidazo[4,5, 1- jk][1,4]benzodiazepin-2(1H)-thione (TIBO R82150)." *Mol Pharmacol* 43(1): 11-6.

Menzo, S., Bagnarelli, P., Giacca, M., Manzin, A., Varalso, P., Clementi,

M. 1992. "Absolute quantitation of viremia in human immunodeficiency virus infection by competitive reverse transcription and polymerase chain reaction." *J Clin Microbiol* 30: 1752-7.

Nunberg, J. H., *et al.* 1991. "Viral resistance to human immunodeficiency virus type 1-specific pyridinone reverse transcriptase inhibitors." *J Virol* 65(9): 4887-92.

Oka, S., Urayama, K., Hirabayaski, Y., Kimura, S., Mitamura, K., Shimada, K. 1992. "Human immunodeficiency virus DNA copies as a virologic marker in a clinical trial with beta-interferon." *J AIDS* 5: 707-11.

Pachl, C., Lindquist, C., Kern, D., *et al.*, June, 1993. "Quantitation of HIV-1 RNA in plasma using a signal amplification branched DNA (bDNA) assay." *IX International Conference on AIDS*, Berlin, WS-A24-1.

Patterson, B., Till, M., Otto, P., *et al.* 1993. "Detection of HIV-1 DNA and messenger RNA in individual cells by PCR-driven in situ hybridization and flow cytometry." *Science* 260: 976-9.

Piatak, M., Luk, K.-C., Williams, B., Lifson, J. 1993. "Quantitative competitive polymerase chain reaction for accurate quantitation of HIV DNA and RNA species." *BioTechniques* 14: 70-80.

Piatek, M., Saag, M., Yang, L., *et al.* 1993. "High levels of HIV-1 in plasma during all stages of infection determined by competitive PCR." *Science* 259: 1749-54.

Poznansky, M., Walker, B., Haseltine, W., Sodroski, J., Langhoff, E. 1991. "A rapid method for quantitating the frequency of peripheral blood cells containing HIV-1 DNA." *J AIDS* 4: 368-73.

Ranki, A., Valle, S., Krohn, M., *et al.* 1987. "Long latency procedes overt seroconversion in sexually transmitted HIV infection." *Lancet* 2: 589-93.

Rasheed, S., Li, Z., Pachl, C., Hoo, B., Todd, J., Kokka, R., June, 1993. "Comparison of the branched DNA (bDNA) signal amplification assay with plasma culture and PCR for the detection and quantitation of HIV-1 RNA in plasma." *IX International Conference on AIDS*, Berlin, PO-B41-2480.

Richman, D. 1993. "Resistance of clinical isolates of human immunodeficiency virus to antiretroviral agents." *Antimicrob Agents Chemother* 37(6): 1207-13.

Richman, D., *et al.* 1991. "BI-RG-587 is active against zidovudine-resistant human immunodeficiency virus type 1 and synergistic with zidovudine." *Antimicrob. Agents Chemother.* 35(2): 305-8.

Richman, D., *et al.* 1991. "Human immunodeficiency virus type 1 mutants resistant to nonnucleoside inhibitors of reverse transcriptase arise in tissue culture." *Proc Natl Acad Sci U S A* 88(24): 11241-5.

Richman, D. D., Guatelli, J. C., Grimes, J., Tsiatis, A., Gingeras, T. 1991. "Detection of mutations associated with zidovudine resistance in human immunodeficiency virus by use of the polymerase chain reaction." *J Infect Dis* 164(6): 1075-81.

Richman, D. D. and the ACTG 164/168 study team, July, 1992. "Loss of nevirapine activity associated with the emergence of resistance in clinical trials." *VIII International Conference on AIDS*, Amsterdam, Abstract POB 3576.

Rogers, M., Ou, C.-Y., Rayfield, M., *et al.* 1989. "Use of the polymerase chain reaction for the early detection of the proviral sequences of human immunodeficiency virus in infants born to seropositive mothers." *N Engl J Med* 320: 1649-54.

Scadden, D., Wang, Z., Groopman, J. 1992. "Quantitation of plasma human immunodeficiency virus type 1 RNA by competitive polymerase chain reaction." *J Infect Dis* 165: 119-23.

Scarlatti, G., Lombardi, V., Plebani, A., *et al.* 1991. "Polymerase chain reaction, virus isolation and antigen assay in HIV-1-antibody-positive mothers and their children." *AIDS* 5: 1173-8.

Schinazi, R. F., *et al.* 1993. "Characterization of human immunodeficiency viruses resistant to oxathiolane-cytosine nucleosides." *Antimicrob Agents Chemother* 37(4): 875-81.

Schnittman, S., Greenhouse, J.J., Psallidopoulos, M.C., *et al.* 1990. "Increasing viral burden in CD4+ T cells from patients with human immunodeficiency virus (HIV) infection reflects rapidly progressive immunosuppression and clinical disease." *Ann Intern Med* 113: 438-443.

Shafer, R. W., *et al.*, June, 1993. "Combination therapy with ZDV + ddI suppresses virus load but does not prevent the emergence of HIV-1 with zidovudine resistance." *IX International Conference on AIDS*, Berlin, Abstract WS-B25-3.

Sheppard, H., Ascher, M., Busch, M., et al. 1991. "A multicenter proficiency trial of gene amplification (PCR) for the detection of HIV-1." *J AIDS* 4: 277-83.

Sheppard, H., Dondero, D., Arnon, J., Winkelstein, W. 1991. "An evaluation of the polymerase chain reaction in HIV-1 seronegative men." *J AIDS* 4: 819-23.

Simmonds, P., Balfe, P., Peutherer, J., Ludlum, C., Bishop, J., Brown, A. 1990. "Human immunodeficiency virus-infected individuals contain provirus in small numbers of peripheral mononuclear cells and at low copy number." *J Virol* 64: 864-72.

St. Clair, M. H., et al. 1991. "Resistance to ddI and sensitivity to AZT induced by a mutation in HIV-1 reverse transcriptase." *Science* 253: 1557-1559.

Telenti, A., Imboden, P., Germann, D. 1992. "Competitive polymerase chain reaction using an internal standard: application to the quantitation of viral DNA." *J Virol Methods* 39(3): 259-68.

Templeton, N. 1992. "The polymerase chain reaction—history, methods, and application." *Diagn Molec Pathol* 1: 58-72.

Tudor-Williams, G., et al. 1992. "HIV-1 sensitivity to zidovudine and clinical outcome in children." *Lancet* 339(8784): 15-9.

Vasudevachari, M. B., et al. 1992. "Prevention of the spread of HIV-1 infection with nonnucleoside reverse transcriptase inhibitors." *Virology* 190(1): 269-77.

Whetsell, A., Drew, J., Milman, G., et al. 1992. "Comparison of three non-radioisotopic polymerase chain reaction-based methods for detection of human immunodeficiency virus type 1." *J Clin Microbiol* 30: 845-53.

Williams, P., Simmonds, P., Yap, P., et al. 1990. "The polymerase chain reaction in the diagnosis of vertically transmitted HIV infection." *AIDS* 4: 393-8.

Winters, M. A., Holodniy, M., Katzenstein, D. A., Merigan, T. C. 1992. "Quantitative RNA and DNA gene amplification can rapidly monitor HIV infection and antiviral activity in cell cultures." *PCR Methods Appl* 1(4): 257-62.

Wolinsky, S., Rinaldo, C., Kwok, S., et al. 1989. "Human immunodeficiency virus type 1 (HIV-1) infection a median of 18 months before a diagnostic Western blot." *Ann Intern Med* 111: 961-72.

Yang, B., Yolken, R., Viscidi, R. 1993. "Quantitative polymerase chain reaction by monitoring enzymatic activity of DNA polymerase." *Anal Biochem* 208: 110-6.

Yen-Lieberman, B., Caret, J., Starkey, C., Spahlinger, T., Reinhardt, C., Jackson, J., October, 1993. "Effects of temperature, time and plasma separation on quantitative plasma HIV-1 RNA levels." *33rd ICAAC Meeting*, New Orleans, Abstract 1258.

CHAPTER 12

Future Directions
of Molecular Pathology

Lawrence M. Silverman and Ruth A. Heim

What is the the future for molecular pathology? In this final chapter we will highlight selected scientific trends and alert the reader to some of their possible implications.

The Human Genome Project

Since the original proposal by Dulbecco in 1986, the Human Genome Initiative (often referred to as the Human Genome Project, which is the NIH-sponsored effort in the USA) has evolved into an international attempt to sequence the genome and to identify and locate human genes. This massive Project, originally headed by James Watson and now by Francis Collins, has sponsored technological advancements that have had an immediate impact on understanding gene structure and function, thus establishing the role of certain genes in genetic disorders and cancer. While the task is formidable, the success of this effort will have profound effects on our ability to provide useful information to society.

The basis of this effort is to construct a detailed "map" of markers that investigators can use as reference points in identifying genes. Several strategies now in use include using cDNA libraries to identify expressed sequences directly. The cDNAs are then mapped to the appropriate chromosomes. Large cloning vectors, e.g., yeast artificial chromosomes or YACs (Anand 1992), are a useful tool for the second level of the Genome Project, which is to clone the entire human genome. The ultimate goal, to identify all human genes and their associated regulatory sequences, will be met by more conventional means (e.g., linkage analysis) as well as by new strategies. One newly described approach is genomic mismatch scanning (Nelson *et al.* 1993), in which large areas of the genome are scanned to identify those regions that are identical-by-descent (usually between related individuals). Where mismatches occur, they can be identified using a microbial enzyme recognition system. Large regions can thus be screened by comparing individuals with family members. While this approach has been successfully tested in yeast, the human genome presents some unique problems that would have to be overcome; for example, the large number of repetitive sequences may give ambiguous, or even false positive, results. However, this approach holds much promise for large-scale mapping projects.

The success of these approaches to identifying genes will complement the currently used techniques of "positional" cloning that have identified genes responsible for cystic fibrosis, Duchenne/Becker muscular dystrophies, neurofibromatosis, fragile X syndrome, myotonic dystrophy, Huntington disease and others, some of which were discussed in previous chapters. Diagnostic testing for these disorders has become available, although there is concern that treatment and cures are not yet available. However, the prospects of rational treatments, even gene therapies, remain theoretically possible. Research into treatments and cures for genetic disorders falls within the purview of the Human Genome Project and is being actively pursued.

Among the first genes identified within the context of the Human Genome Project to have a potentially significant impact will probably be a human breast cancer gene (BRCA1), already mapped to chromosome 17. This gene may be responsible for as much as 5% of all breast cancer and a significant proportion of ovarian cancer (Easton *et al.* 1993). Clinical laboratories will potentially be able to identify individuals at risk for developing breast cancer, an unprecedented form of susceptibility testing. Moreover, BRCA1 is potentially one of the first candidate genes for population screening, since huge cost-savings and preventative treatment may be possible. Other Human Genome Project-sponsored research with similar impact includes identification of genes for colon cancer (Fishal *et al.* 1993) and MODY (maturity-onset diabetes of the young).

Gene Therapy

As the Human Genome Project progresses and new associations between genes and specific disorders are found, the field of molecular diagnostic testing will have to keep pace. The development of automated technology and the availability of highly trained technical personnel will maintain standards of diagnostic care, but the ability to improve the quality of life for patients with diagnosed disorders may depend in part upon developments in gene therapy. All requests for gene therapy are currently reviewed by the Recombinant DNA Advisory Committee (RAC) and the Food and Drug Administration. For more detailed accounts of gene therapies and their implications, readers are referred to recent reviews, including Miller (1992), Morgan and Anderson (1993), Mulligan (1993), and Ledley (1993).

Gene therapies can fall into the following specific categories: (1) protein modification therapy, perhaps involving recombinant technology, (2) somatic gene replacement or correction, and (3) germline gene therapy.

Protein Modification Therapy

A number of disorders can be treated by supplying the missing or defective protein, or dietary restriction can reduce the intake of substances which cannot be metabolized, usually due to defective or missing enzymes. For example, extracellular replacement using recombinant factor VIII can be used to treat hemophilia A patients who are deficient in this coagulation protein. One advantage of preparing recombinant proteins is their high purity due to lack of contaminants associated with previous purification procedures, as was the case when factor VIII was prepared from blood donors, some of whom were HIV-positive. Recently, gene therapy was successfully used to correct clotting defects in a canine model of hemophilia B (Kay *et al.* 1993).

In a similar vein, children who are immunodeficient due to the absence of active adenosine deaminase (ADA), a key enzyme in purine metabolism, can be treated by receiving intravenous ADA, chemically modified to increase its circulating half-life (Hershfield *et al.* 1993). Or alpha1-antitrypsin, a protease inhibitor intrinsic to neutralizing neutrophil elastase and collagenase in the lung, can be administered intravenously to ameliorate the effects of alpha1-antitrypsin deficiency (Sandhaus 1993).

Somatic Gene Replacement

Replacing defective or missing proteins is only useful in a few disorders, in which a small amount of replacement protein can ameliorate symptoms and the location of the replacement is easily accessible. For most genetic disorders, a more rational approach is to provide a continuous supply of the appropriate protein by introducing genes into vectors that can then be transferred into cells which express the gene product. This has been accomplished using viral vectors for Duchenne muscular dystrophy, adenosine deaminase deficiency, alpha1-antitrypsin deficiency, cystic fibrosis and other disorders. Results are promising, although considerable work remains to be done before these approaches become a therapeutic possibility.

Germ-line Gene Therapy

Unlike somatic gene therapy, germ-line gene therapy involves correcting or preventing gene defects by transferring appropriate genes into reproductive cells (Wivel and Waters 1993). This kind of gene therapy raises more issues than others because of the potential to transmit traits to future generations. The development of transgenic animal models and the micromanipulation of embryos (see below) suggest that germ-line therapy will be technically feasible. However, scientific and ethical issues remain to be resolved before this exciting area can be fully developed or accepted.

From a scientific perspective, transgenic animal models have become an important tool in studying the biology of numerous genetic diseases, not to mention other mechanisms involved in normal development (Smithies 1993). Transgenic animals are produced by manipulating the DNA in embryonic stem cells, usually in mice. The stem cells differentiate into both somatic and germ-line tissue; hence the transgenic mice have essentially received "germ-line" therapy. The technology cannot be used in humans because the equivalent of the embryonic stem cell has not been identified. Some examples of transgenic models of genetic diseases include Lesch-Nyhan syndrome, Duchenne muscular dystrophy, sickle cell anemia, Gaucher's disease, cystic fibrosis, familial hypercholesterolemia and other, less familiar disorders.

It is beyond the scope of this book to delve into the many unresolved ethical considerations regarding germ-line gene therapy. A few general issues include the following: (1) allocation of resources (i.e., expense associated with the procedures); (2) risks associated with irreversible germline changes; and, (3) bioethics of "improving the species." Some issues may not be resolvable, but guidelines may be laid down during the next few years of discussions between groups of scientists, ethicists, politicians, lawyers and other concerned citizens. The Human Genome Project is funding studies of these issues.

Genes and Patents

One of the more controversial aspects of the new molecular biotechnology involves rights to discoveries, payments to patent holders and the nature of what are proprietary rights in this area. For instance, is DNA sequence the property of the individual who first reported it even if the sequence is not known to define a functioning gene? Questions such as this are being posed both to regulatory agencies and to scientific organizations.

Ethical, Legal and Social Issues

Ethical, legal and social issues identified by the Human Genome Project include fairness in the use of genetic information, the impact of such information on individuals, privacy and confidentiality issues, the impact on genetic counseling, the impact on reproductive decisions (and the associate questions of what constitutes a "mild" disease versus a trait), issues raised by the introduction of genetics into mainstream medical practice, uses and misuses of genetics in the past, questions raised by commercialization, and conceptual and philosophical implications. Readers interested in these issues are referred to chapter 9 and to Annas and Elias (1992).

Personnel

As more genes are associated with specific disorders and as our ability to act upon diagnostic information increases, the number of genetic tests performed will definitely increase. However, we will still face the inevitable concerns about cost-effective utilization of limited laboratory resources. Eventually, automation will decrease cost per test, and regulatory agencies will dictate requirements for quality assurance programs. Inevitably, the limiting factor for expansion of services will be the availability of highly trained personnel to perform and interpret these analyses. Where will these individuals come from?

Although regulatory agencies, both state and federal, are intimately involved with clinical laboratories, it is presumptive to predict the future of health care; however, certain analogies can be drawn from the current situation in clinical laboratories. Medical technologists, also referred to as clinical laboratory scientists, are highly trained professionals with extensive experience in delivering laboratory data using "state-of-the-art" technology. These laboratorians are certified through experience and board examination by the American Society of Clinical Pathologists. In our Molecular Genetics Laboratory at the University of North Carolina Hospitals, clinical laboratory scientists provide the technical and supervisory skills, in addition to the constant research and development necessary to keep abreast of new technological breakthroughs. Directors of molecular laboratories, who interpret results and interact with other health care professionals, will also have to meet regulatory agencies' requirements; currently, the American Board of Medical Genetics provides board certification in Molecular Genetics.

The information generated from a molecular genetics laboratory is only meaningful if placed in the appropriate context and acted upon by both the physician and the patient in a timely manner. Too often, the generation and delivery of data represent the perceived end of the laboratory's role; however, helping families with genetic disorders should involve a team of health care professionals whose roles include interpretation and integration of laboratory and clinical information so that informed decisions can be made. Key individuals in this process are genetic counselors, who are also certified by the American Board of Medical Genetics. In our institution, as in most other medical centers, all requests for molecular genetic family studies are referred to the genetics counseling service, where decisions can be made about the appropriate utilization of laboratory services. If laboratory testing is appropriate, the data are disseminated to the ordering physician, along with individual or family counseling (if desired). This assumes that medical schools and continuing medical education programs will provide the necessary education and information to help physicians screen genetic testing referrals due to the current lack of genetic professionals.

Considering that the clinical laboratory may be providing presymptomatic or susceptibility testing in the near future (for example, for the

BRCA1 gene associated with breast cancer or the genes associated with colorectal cancer), there are not enough trained professionals to address the needs of the at-risk families and individuals. As broad-based (voluntary) population screening becomes technically feasible and, perhaps, desirable, personnel issues will become critical. For instance, general practitioners could face the possibility of not having sufficient information for patients to make rational and voluntary decisions regarding these tests. At present, counseling is frequently provided to pregnant women, but often this is not timely and negates many options. Future counseling needs should be addressed now, because the potential of knowing genetic susceptibilities will have an impact on every physician and patient.

Technology

In the preceding chapters, current technologies associated with particular analyses were described, but novel approaches to identifying acquired and inherited genetics changes are still being initiated. Here we present a sampling of new and developing techniques that have an impact on molecular diagnostic testing.

Mutation Detection Using DNA

Approaches to mutation-detection schemes, methods which screen stretches of nucleotide sequence for variations, are constantly being refined. For a variety of genetic disorders, having to detect multiple disease-causing mutations within the same genetic locus, because of allelic heterogeneity, makes laboratory analysis cumbersome and inefficient (for example, see chapter 6). Being able to screen 100's or 1000's of nucleotides rapidly and to determine whether any sequence variation is disease-causing would represent a significant improvement over current approaches that identify individual known mutations or rely on pooled approaches (where multiple mutations are screened for, but not specifically identified, by pooling techniques).

The ideal scheme would be effective in detecting single base changes, as well as larger deletions, insertions or rearrangements. Moreover, the procedure should be relatively easy, inexpensive and rapid. At the present time, no single approach for mutation screening is sufficient; usually, a combination of approaches must be used to effectively screen for mutations, including single-stranded conformational analysis or SSCA, denaturing gradient gel electrophoresis, chemical cleavage and heteroduplex analysis. Using one or a combination of these methods, sequence variants are identified and can only be fully characterized by sequencing. Computer-assisted comparisons with known sequence can determine the nature of the sequence variation, and data-base reviews of reported mutations can ascertain whether the variation is novel or previously reported. Newer techniques that are less expensive and less tedious will have to be developed to gain wide acceptance in clinical laboratories. A general approach will probably have tremendous impact on diagnosing genetic disorders and genes altered in neoplasia, most of which are characterized by extensive allelic heterogeneity.

One mutation-detection scheme that may be of use clinically is a modification of standard sequencing protocols designed for identifying specific mutations, called solid-phase *minisequencing* (Syvanen *et al.* 1990). In this technique, specific oligonucleotide primers define a sequence into which nucleotides are incorporated. Labeled nucleotides that correspond to the normal (wild-type) sequence can then be distinguished from unlabeled nucleotides representing single base mutations. This entire process can be automated using various robotic processors, making it an attractive procedure for a diagnostic laboratory.

Mutation Detection Using RNA

A commonly used extension of previous techniques is mRNA-reverse transcriptase-PCR (RT-PCR), which can detect the presence of certain variants that alter transcribed mRNA and potentially their encoded proteins. Key to this approach is the requirement of preserving the integrity of the mRNA, something which is not preserved in the usual sampling procedures used to obtain genomic DNA. Another consideration is the choice of tissue—most genes are only expressed in certain tissues. Thus, for the study of a gene which is only expressed in muscle (such as dystrophin), muscle tissue must be the source of the mRNA. However, use of PCR has allowed investigators to take advantage of the low level of expression which occurs in all tissues (called "illegitimate" expression) to amplify cDNAs from small amounts of mRNA.

Following RT-PCR, the cDNA product can be analyzed as mentioned above or can be used as a template for *in vitro* synthesis of proteins. This method has been called the *protein truncation test* (PTT) (Roest *et al.* 1993), since mutations which result in shortened or truncated proteins can be detected. For example, deletions in the dystrophin gene (see chapter 7) would result in shortened proteins. Applications of the technology include analysis of the altered gene product observed in familial adenomatous polyposis (Powell *et al.* 1993).

A newly described mRNA-based approach is called *differential display* (Liang and Pardee 1992), in which the total transcribed message from one tissue is assessed and compared, either to message from other tissues or from other individuals. This is accomplished by using reverse transcription and random-priming PCR to produce an array of cDNA bands after electrophoresis. Bands are identified with an appropriate probe, usually after Southern blotting. Products unique to one set of bands are then sequenced. This strategy has been used to evaluate malignant tissue and identify specifically altered cDNAs, such as those associated with the APC gene in familial adenomatous polyposis (Horii *et al.* 1993).

Diagnostic Testing in Fetal Cells

Since much genetic testing involves prenatal diagnosis, techniques for sampling fetal tissue are important. In addition to traditional techniques, such as amniocentesis and chorionic villus sampling, the isolation of fetal cells from maternal circulation promises to be an exciting source of fetal nucleated red blood cells (Holzgreve *et al.* 1992). Specific receptors on the surface of these cells allow them to be isolated or sorted using flow cell cytometry. The cells can be analyzed by FISH or standard PCR techniques for various chromosomal or genomic alterations. However, recent reports of the presence of cells from previous pregnancies that persist in the maternal circulation may present a serious drawback to be overcome.

Recently, techniques have been developed that allow genetic analyses to be performed on both oocytes and embryos for the purposes of in vitro fertilization or selected implantation (thus the designation, "preimplantation" genetics) (reviewed in Simpson 1992). These techniques are most suitable for couples who have been identified as carriers of a specific genetic disorder, either through testing or through a previously affected offspring. The procedures are only performed at a few institutions at great expense, and the techniques are still being developed and discussed by social scientists, ethicists and health care professionals.

Conclusions

During the past ten years, the area of molecular pathology has evolved due to exciting technological developments that were applied to clinical disciplines, including genetics, infectious disease and oncology. As we enter the next ten years, expectations are high that improvements in technology and instrumentation will increase useful information at a reasonable cost to the patients and third-party payers. But expectations are just as high that exciting developments emanating from research in gene therapy and the Human Genome Project will challenge clinical laboratorians to become even more involved with efforts to not only diagnose diseases at the molecular level but to actually alter outcome. With these challenges come the realities of cost containment, resource allocation, personnel training, education, and the serious discussions surrounding the various ethical and social issues previously mentioned. We have attempted to describe our efforts at the University of North Carolina as a "snapshot" of the present; we anticipate future developments in this field and hope the next ten years are as eventful as the past.

References

Anand, R. 1992. Yeast artificial chromosomes (YACs) and the analysis of complex genomes. *TIBTECH* 10, 35-40.

Annas, G.J., Elias, S., Eds. 1992. *Gene Mapping Using Law and Ethics as Guides.* Oxford University Press, New York.

Easton, D.F., Bishop, D.T., Ford, D., Crockford, G.P., and the Breast Cancer Linkage Consortium. 1993. Genetic linkage analysis in familial breast and ovarian cancer: results from 214 families. *Am. J. Hum. Genet.* 52, 678-701.

Fishel, R., Lescoe, M.K., Rao, M.R., *et al.* 1993. The human mutator gene homolog MSH2 and its association with hereditary nonpolyposis colon cancer. *Cell* 75(5), 1027-38

Hershfield, M.S., Chaffee, S., Sorenson, R.U. 1993. Enzyme replacement therapy with polyethylene glycol-adenosine deaminase in adenosine deaminase deficiency: Overview and case reports of three patients, including two now receiving gene therapy. *Pediatr. Res.* 33, 42-48.

Holzgreve, W., Garritsen, H.S., Ganshirt-Ahlert, D. 1992. Fetal cells in the maternal circulation. *J. Reprod. Med.* 37, 410-418.

Horii, A., Nakatsuru, S., Ichii, S., Nagase, H., Nakamura, Y. 1993. Multiple forms of the APC gene transcripts and their tissue-specific expression. *Hum Molec Genet* 2, 283-287.

Kay, M.A., Rothenberg, S., Landen, C.N., *et al.* 1993. In vivo gene therapy of hemophilia B: Sustained partial correction in factor IX-deficient dogs. *Science* 262, 117-119.

Liang, P., Pardee, A.B. 1992. Differential display of eukaryotic messenger RNA by means of the polymerase chain reaction. *Science* 257, 967-971.

Ledley, F-D. 1993. Are contemporary methods for somatic gene therapy suitable for clinical applications? *Clin. Invest. Med.* 16, 78-88.

Miller, D.A. 1992. Human gene therapy comes of age. *Nature* 357, 455-460.

Morgan, R.A., Anderson, W.F. 1993. Human Gene Therapy. *Annu. Rev. Biochem.* 62,191-217.

Mulligan, R.C. 1993. The basic science of gene therapy. *Science* 260, 926-932.

Nelson, S.F., McCusker, J.H., Sander, M.A., Kee, Y., Modrich, P., Brown, P.O. 1993. Genomic Mismatch scanning: A new approach to genetic linkage mapping? *Nat. Genet.* 4(1), 5-6.

Powell, S.M., Peterson, G.M., Krush, A.J., *et al.* 1993. Molecular diagnosis of familial adenomatous polyposis. *New Engl J Med* 329(27), 1982-87.

Roest, P.A.M., Roberts, R.G., Sugino, S. , van Ommen, G-J.B., den Dunnen, J.T. 1993. Protein truncation test (PTT) for rapid detection of translation-terminating mutations. *Human Molec Genet* 2(10), 1719-21.

Smithies, O. 1993. Animal models of human genetic diseases. *Trends. Genet.* 9, 112-116.

Sandhaus, R.A. 1993. Alpha 1-antitrypsin augmentation therapy. *Agents Actions Suppl.* 42, 97-102.

Simpson, J.L. 1992. Preimplantation genetics and recovery of fetal cells from maternal blood. *Curr. Opin. Obstet. Gynecol.* 4, 295-301.

Syvanen, A.C., Aalto-Setala, K., Harju, L., Kontula, K., Soderlund, H. 1990. A primer-guided nucleotide incorporation assay in the genotyping of apolipoprotein E. *Genomics*, 8, 684-692.

Wivel, N.A., Waters, L. 1993. Germ-line gene modification and disease prevention: some medical and ethical perspectives. Science, 262, 533-537.

Index

N